Novelties in Schiff Bases

Edited by Takashiro Akitsu

Published in London, United Kingdom

Novelties in Schiff Bases
http://dx.doi.org/10.5772/intechopen.114475
Edited by Takashiro Akitsu

Contributors
Arif Ullah, Bredan Chidozie Asogwa, Chinedum Ifeanyi Nwankwo, Daisuke Nakane, Elias Emeka Elemike, Felix Chigozie Nwadire, Ifeanyi Edozie Otuokere, Jayarama Shankar, Jyotiprabha Rout, Kareem Shah Bukhari, Malavalli Guruswamy Dileep Kumar, Mamoun M. Bader, Md. Abdul Hannan, Md. Hasibul Islam, Mohammed Umar Adaji, Moses Saviour Iorungwa, Nabakrushna Behera, Olalekan Wasiu Salawu, Onyinyechi Uloma Akoh, Phuong-Truc T. Pham, Precious Onyinyechi Emole, Rekha Barku More, Rin Tsuchiya, Sasmita Moharana, Savita V. Thakare, Shilpa Laxman Sangle, Takashiro Akitsu, Tankadhar Behera, Yuya Higashi

First published in London, United Kingdom, 2024 by IntechOpen
IntechOpen is the global imprint of INTECHOPEN LIMITED, registered in England and Wales, registration number: 11086078, 167-169 Great Portland Street, London, W1W 5PF, United Kingdom

British Library Cataloguing-in-Publication Data
A catalogue record for this book is available from the British Library

Additional hard and PDF copies can be obtained from orders@intechopen.com

Novelties in Schiff Bases
Edited by Takashiro Akitsu
p. cm.
Print ISBN 978-0-85014-072-9
Online ISBN 978-0-85014-073-6
eBook (PDF) ISBN 978-0-85014-074-3

For EU product safety concerns:
IN TECH d.o.o., Prolaz Marije Krucifikse Kozulić 3, 51000 Rijeka, Croatia,
info@intechopen.com or visit our website at intechopen.com.

Meet the editor

Takashiro Akitsu has been a Professor of Chemistry at the Department of Chemistry, Faculty of Science, Tokyo University of Science, Japan, since 2016. He graduated from the Department of Chemistry, Osaka University, Japan, in 1995 and obtained his Ph.D. from the same university in 2000. Dr. Akitsu studied at the Institute for Protein Research, Osaka University, and moved to the Department of Chemistry, Keio University, Japan. He visited the Department of Chemistry at Stanford University, USA, before moving to Tokyo University of Science. His research field is inorganic coordination chemistry and crystallography. His research interest is hybrid materials of Schiff base metal complexes.

Contents

Preface

Schiff bases, first reported by H. Schiff, are long-known organic compounds containing imine or azomethine (RR'C=NR") groups obtained from the dehydration condensation of aldehydes and primary amines. Schiff bases have long been known not only as compounds but also as typical ligands for metal complexes, as they are Lewis bases that can donate lone pairs of electrons on nitrogen atoms. Although Schiff bases can be said to be classic and typical, they can also be said to be evergreen compounds for which new research results are still actively being reported. This book introduces novel aspects of Schiff bases.

For example, the authors (editors) have systematically studied amino acid Schiff base copper complexes. The skeleton is a compound that has been known for a long time, and there are approximately 20 types of L-α amino acids that can be substituted. For molecular design, we varied the substituents to be introduced into aldehydes, introduced azobenzene groups, and chemically modified organic ligands, such as esterification of amino acids. Synthesis methods are diverse, including one-pot synthesis in a conventional solution, acid-base control, and solvent selection. Additionally, my coworkers and I gained experience with mechanochemical synthesis, microwave synthesis, and microfluidic device synthesis. In terms of physical properties and functions, in addition to those derived from chirality, we report reactions with photocatalysts, hexavalent chromium reduction, SOD activity, antibacterial activity, and so on. Furthermore, the complexes were not only used alone but also adsorbed on metal

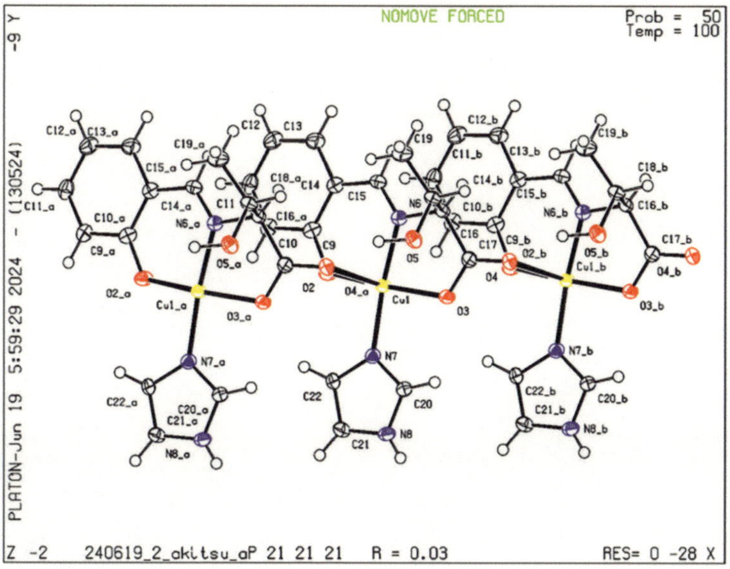

Figure 1.
Crystal structure of L-threonine derivative Schiff base copper complex.

oxide surfaces, bound to proteins, and used as constituent elements of nanoscale composite materials.

Evolution of analytical methods is also a driving force for incorporating additional perspectives. We performed X-ray crystal structure analysis of the L-threonine derivative Schiff base copper complex whose structure is undetermined (after checking with a crystal structure database) and investigated intermolecular interactions as a crystal using Hirshfeld surface analysis, which we have reported. Reanalysis of previously reported crystal structures using an X-ray diffractometer capable of producing microcrystals led to the acquisition of new statistics, such as a comparison of the accuracy of data under severe conditions (**Figure 1**, analyzed by Dr. Yukiyasu Kashiwagi, ORIST, Japan).

In the secondary use of measurement data acquired in the future, a "data management plan" targeting various analyses may become more common, and Schiff bases may receive less attention as a group of compounds with a large amount of big data and accumulation.

Takashiro Akitsu
Faculty of Science,
Department of Chemistry,
Tokyo University of Science,
Tokyo, Japan

Chapter 1

Schiff Bases: Contemporary Synthesis, Properties, and Applications

Md. Hasibul Islam and Md. Abdul Hannan

Abstract

Schiff bases, a versatile class of organic compounds with a (>C═N−) functional group, have gained significant attention in contemporary chemistry due to their remarkable synthesis methods, diverse properties, and wide-ranging applications. This book chapter focused on Schiff bases, emphasizing their versatility in contemporary chemistry. It covers synthesis methods, structural aspects, and applications across multiple domains. The historical development of Schiff bases is traced, emphasizing their pivotal role in modern chemistry as key intermediates in various processes. The chapter delves into synthesis methods, including traditional, sustainable, and green chemistry approaches, and discusses structural aspects, spectroscopic characterization, thermodynamic properties, and kinetics. Schiff bases are categorized into various types, highlighting their diversity. Applications are explored in catalysis, asymmetric synthesis, coordination chemistry, medicinal chemistry, materials science, and environmental contexts. The chapter concludes with recent advances, emerging trends, and future directions, positioning it as a valuable resource for researchers, students, and practitioners interested in Schiff bases. The proposed chapter aims to explore the latest developments in Schiff bases, providing an in-depth review of their synthesis, properties, and applications across various scientific fields.

Keywords: Schiff bases, metal complexes, characterization, synthesis, properties, applications

1. Introduction

1.1 Definition and historical background of Schiff bases

1.1.1 Definition

The term "Schiff bases" refers to a class of organic compounds resulting from the condensation reaction between a primary amine and a carbonyl compound, typically an aldehyde or a ketone. Named after the German chemist Hugo Schiff, who first reported their synthesis in 1864 [1], these compounds have since become integral to the fabric of organic chemistry.

Schiff bases have significantly influenced the advancement of coordination chemistry, playing a pivotal role in the development of inorganic and bioinorganic chemistry, as well as optical materials. Their widespread use extends across various branches of chemistry, including inorganic, organic, and analytical fields, making them a significant component of commonly employed organic molecules.

Schiff bases are characterized by a central carbon-nitrogen double bond, also known as an imine or azomethine group, formed during the reaction. The general structure can be represented as $R_1R_2C{=}NR_3$, where R_1, R_2, and R_3 denote organic substituents.

In coordination chemistry, Schiff bases are common ligands that are classified based on the number of donor atoms, including uni-, di-, tri-, and tetra-dentate ligands, and are derived from aromatic aldehydes and alkyl diamines (**Figures 1** and **2**) [2].

1.1.2 Historical background

The historical trajectory of Schiff bases is intricately linked with the evolution of organic chemistry in the nineteenth century. Hugo Schiff's pioneering work marked the initial strides into the synthesis and exploration of these compounds [3].

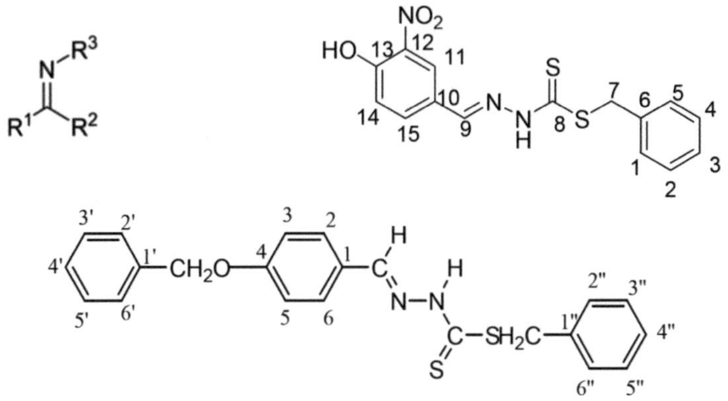

Figure 1.
General structure of Schiff bases and few examples of Schiff bases.

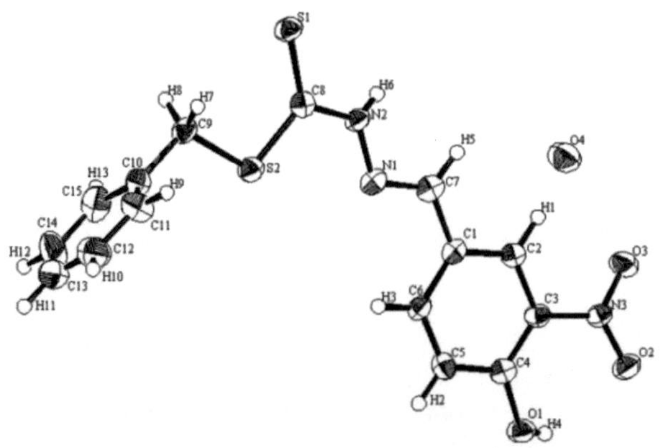

Figure 2.
ORTEP diagram (thermal ellipsoids at 40% probability level) of S-benzyl-β-N-(4-hydroxy-3-nitrophenyl) methylenedithiocarbazate with atom numbering scheme.

Schiff himself discovered the condensation reaction while investigating the reactions of amines with aldehydes and ketones. In the decades following Schiff's groundbreaking discovery, chemists delved deeper into understanding the underlying principles governing Schiff bases. Noteworthy contributions came from eminent figures such as Emil Fischer, who not only expanded the scope of Schiff base reactions but also provided insights into their stereochemistry [4]. The historical significance of Schiff bases extends beyond their fundamental role in chemical reactions; they played a crucial role in advancing the broader field of organic synthesis. The synthesis of these compounds has undergone refinement and diversification over time, resulting in a broad array of derivatives with varying properties and applications.

The Schiff bases derived from the dithiocarbazates, thiosemicarbazides, and dithiocarbamates form an interesting series of compounds whose properties can be greatly modified by introducing organic substituents into the molecule thereby causing variation in the ultimate donor properties and inducing different stereochemistry in the resultant metal complexes. *S*-Benzyl dithiocarbazate (SBDTC) Schiff bases comprise an essential part of nitrogen-sulfur donor ligands [5, 6].

Since the synthesis of dithiocarbamic acids by Curtius more than a century ago, a significant number of S-alkyl/aryl-esters, their related complexes, and the Schiff bases of these compounds have been developed for potential use in antibacterial, anticancer, antifungal, antitumor, antiamoebic therapy, anti-inhibitory cell migration, analgesic and anti-inflammatory activities, or merely for their diverse coordination geometries [7–10].

Synthetic chemists extensively utilize Schiff bases and related complexes for diverse processes, ranging from the oxidation of alkenes to the catalytic transformation of hydrocarbons. These processes yield valuable oxygenated derivatives such as alcohols, aldehydes, and epoxides.

As we navigate the historical landscape of Schiff bases, it is essential to recognize how their discovery marked a pivotal moment, setting the stage for continuous exploration and contributing to the rich tapestry of organic chemistry. Today, the historical journey of Schiff bases serves as the foundation for the contemporary understanding and applications explored in subsequent sections of this chapter.

1.2 Significance of Schiff bases in modern chemistry

In the vast and dynamic realm of modern chemistry, Schiff bases have emerged as compounds of profound significance, playing pivotal roles in various facets of scientific exploration. Their versatility, reactivity, and distinctive properties have positioned them as indispensable building blocks across diverse fields.

Schiff bases exhibit a diverse range of biological activities, including antibacterial, antimycobacterial, antifungal, anti-inflammatory, antioxidant, anticancer, etc. [11–14]. Initially discovered by Hugo Schiff in the nineteenth century, these compounds have evolved to become integral components of contemporary research and synthesis. Schiff bases find extensive use in coordination chemistry, catalysis, and medicinal chemistry. Their ability to form stable complexes with metal ions has led to their widespread adoption in the design of novel catalysts and coordination compounds [15]. Additionally, Schiff bases serve as crucial intermediates in the synthesis of various pharmaceuticals, agrochemicals, and dyes, contributing significantly to the development of new therapeutic agents [16]. In the realm of materials science, Schiff bases play a key role in the design and fabrication of functional materials. Their unique electronic and optical properties make them valuable in the development of sensors, molecular switches, and advanced materials for electronic devices [17].

Furthermore, the use of Schiff bases in supramolecular chemistry has opened avenues for the creation of self-assembled structures and molecular machines, showcasing their importance in the nanotechnology field [18]. The navigation of the complexities of modern chemistry and the enduring relevance of Schiff bases is a testament to their adaptability and multifaceted contributions to scientific advancements.

1.3 Overview of the chapter's content

This chapter is a comprehensive exploration of Schiff bases, spanning their historical roots to their contemporary significance in modern chemistry. The content is structured to provide a thorough understanding of the synthesis, properties, and applications of Schiff bases, highlighting their diverse roles across various scientific domains.

1. *Introduction to Schiff bases*: the journey begins with a detailed introduction, defining Schiff bases and tracing their historical background. We revisit the foundational work of Hugo Schiff and the subsequent developments that have shaped our understanding of these compounds.

2. *Synthesis of Schiff bases*: the chapter then delves into the contemporary methods of synthesizing Schiff bases. From traditional condensation reactions to innovative strategies, this section explores the diverse synthetic routes that enable the creation of Schiff bases with tailored structures and functionalities.

3. *Properties of Schiff bases*: a critical examination of the unique properties exhibited by Schiff bases follows the synthesis section. Their electronic configuration, structural flexibility, and other distinctive characteristics are dissected to provide a nuanced understanding of how these properties influence their behavior in various chemical contexts.

4. *Catalytic and synthetic applications*: Schiff bases have found wide-ranging applications in catalysis and synthetic methodologies. This section investigates their catalytic prowess and their role in facilitating diverse transformations, showcasing how they contribute to the advancement of synthetic chemistry.

5. *Medicinal chemistry and biological activities*: the discussion then shifts to the medicinal chemistry aspects of Schiff bases. Their significant impact on drug design and development is explored, emphasizing their potential as therapeutic agents with antibacterial, antifungal, antiviral, and anticancer properties.

6. *Materials science and coordination chemistry*: Schiff bases play a central role in materials science and coordination chemistry. This section illuminates their contributions to the design of functional materials, including coordination polymers and metal-organic frameworks, with applications ranging from sensors to catalytic supports.

7. *Supramolecular chemistry and analytical applications*: the chapter concludes by examining the involvement of Schiff bases in supramolecular chemistry and their utility in analytical applications. Their role in the construction of supramolecular assemblies and their use as fluorescent probes and sensors for chemical detection are explored.

Throughout this journey, we aim to provide readers with a comprehensive and insightful overview of Schiff bases, showcasing their historical significance, contemporary synthesis methods, unique properties, and diverse applications. By the chapter's end, readers will gain a profound appreciation for the multifaceted nature of Schiff bases and their enduring impact on the landscape of modern chemistry.

2. Synthesis of Schiff bases

The synthesis of Schiff bases has been the subject of extensive investigation, with diverse methodologies explored to obtain Schiff bases featuring varied structures and functionalities. Numerous studies have been conducted to undertake the synthesis of Schiff bases [19].

Historically, the condensation reaction between primary amines and aldehydes or ketones has been a cornerstone in Schiff base synthesis. Pioneered by Hugo Schiff, this classical approach involves the nucleophilic attack of the amino group on the electrophilic carbon of the carbonyl group, resulting in the characteristic carbon-nitrogen double bond. Contemporary methods in Schiff base synthesis have progressed to incorporate inventive strategies, leading to enhanced reaction rates, increased yields, and diminished environmental impact.

The synthesis of Schiff bases encompasses a diverse array of methods that have evolved over time.

2.1 Traditional methods for Schiff base formation

The synthesis of Schiff bases, a class of organic compounds with diverse applications, has a rich historical foundation rooted in traditional methods. Historically, the condensation reaction between primary amines and aldehydes or ketones has been the cornerstone of Schiff base formation. This classical approach, pioneered by Hugo Schiff himself, has provided a fundamental understanding of the reaction principles and laid the groundwork for further exploration [1, 20].

The traditional method of Schiff base formation has been extensively studied and documented. Classic examples include the reaction between aromatic amines and aldehydes. For instance, the condensation of aniline with benzaldehyde leads to the formation of the Schiff base known as Schiff's base (**Figure 3**).

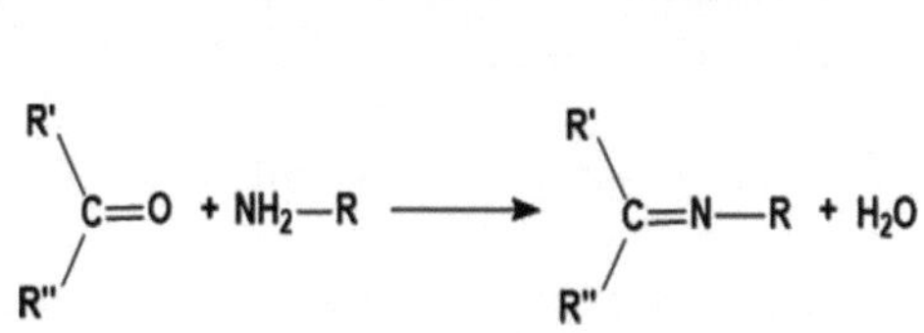

Figure 3.
General synthesis reaction of Schiff base.

Example:

$$C_6H_5CHO + C_6H_5NH_2 \rightarrow C_6H_5CH=NC_6H_5 + H_2O$$

This foundational method has been applied to a broad range of amine and carbonyl reactants, contributing to the synthesis of numerous Schiff bases with diverse structural motifs.

2.2 Modern approaches and green chemistry aspects in Schiff base synthesis

The synthesis of Schiff bases has undergone a transformative evolution with the integration of modern approaches and green chemistry principles. Embracing green chemistry, these modern approaches prioritize sustainable practices by utilizing eco-friendly solvents, such as water, and employing catalysis with recyclable metal complexes or enzymes, which reduces the environmental footprint of the synthesis process. Microwave-assisted synthesis has gained prominence as an energy-efficient method for accelerating chemical reactions, including Schiff base formation. This approach not only reduces reaction times but also enhances overall efficiency, aligning with the principles of green chemistry. Modern approaches often seek metal-free catalysis or the use of eco-compatible metal catalysts, ensuring a more sustainable and greener synthetic process [21, 22]. The reference to green chemistry aspects is exemplified by the work of Anastas and Warner in "Green Chemistry: Theory and Practice" [23]. Green synthesis methods prioritize sustainability, minimizing environmental impact and resource consumption. Several innovative strategies have emerged to enhance the eco-friendliness of Schiff base synthesis.

These modern strategies are reflected in recent literature and contribute to the ongoing advancement of Schiff base synthesis with a focus on sustainability and environmental responsibility.

2.3 Catalysts and conditions for efficient Schiff base formation

Efficient formation of Schiff bases, a pivotal process in organic synthesis, relies on judicious selection of catalysts and reaction conditions. Traditional methods often involve metal catalysts, but contemporary approaches prioritize eco-friendly alternatives to mitigate environmental impact. Metal-free catalysis has emerged as a prominent strategy for green and sustainable Schiff base synthesis. Organic or bio-inspired catalysts are preferred due to their effectiveness and reduced environmental footprint. The solvent can impact reaction rates, selectivity, and the overall sustainability of the synthetic process.

Additionally, the optimization of reaction conditions plays a crucial role. Mild conditions, such as ambient temperature and environmentally benign solvents, contribute to the efficiency and eco-friendliness of Schiff base formation.

2.4 Preparation methods of some important recent synthesized Schiff bases

The synthesis of Schiff bases encompasses a historical journey rooted in classical methods and a contemporary evolution marked by innovative strategies. However, some important Schiff bases have been recently synthesized, which are described here.

2.4.1 1,2-bis((benzylsulfanyl){2-[1-(2-hydroxyphenyl)ethylidene] hydrazin-1-ylidene}-methyl)disulfane

The title compound consists of two Schiff base moieties, namely two *S*-benzyl-*β*-*N*-(2-hydroxyphenylethylidene)dithiocarbazate groups, which are connected through an S—S single bond. These two moieties are twisted with respect to each other, with a dihedral angle of 87.88 (4)° between the S2C N planes (**Figure 4**).

2.4.1.1 Synthesis and crystallization

The ligand precursor, *S*-benzyl dithiocarbazate (SBDTC), was prepared using the literature method [20]. The title compound was prepared as follows: to the ligand precursor, SBDTC (0.99 g, 5 mmol) dissolved in ethanol (40 ml) was added 2-hydroxy acetophenone (0.68 g, 5 mmol), and the aliquot was heated under reflux for 1 h. The resultant yellow solution was cooled to room temperature. The light-yellow precipitate that formed was filtered off, washed with hot ethanol, and dried under vacuum over anhydrous $CaCl_2$ (yield: 1.23 g, 73.65%). The prepared compound (0.17 g) was dissolved in acetonitrile (20 ml) on warming and mixed with ethanol (10 ml). Light-yellow platelet single crystals of the title compound (m.p. 386–387 K) suitable for X-ray study were obtained after 17 days, along with colorless needle-shaped crystalline solids (m.p. 413–418 K) (**Figure 5**) [24].

2.4.2 Synthesis of the Schiff base S-benzyl-β-N-(4-benzyloxyphenyl) methylenedithiocarbazate

The reaction of 4-benzyloxybenzaldehyde with *S*-benzyl dithiocarbazate (SBDTC) afforded the Schiff base *S*-benzyl-*β*-*N*-(4-benzyloxyphenyl)methylenedithiocarbazate (**Figure 6**).

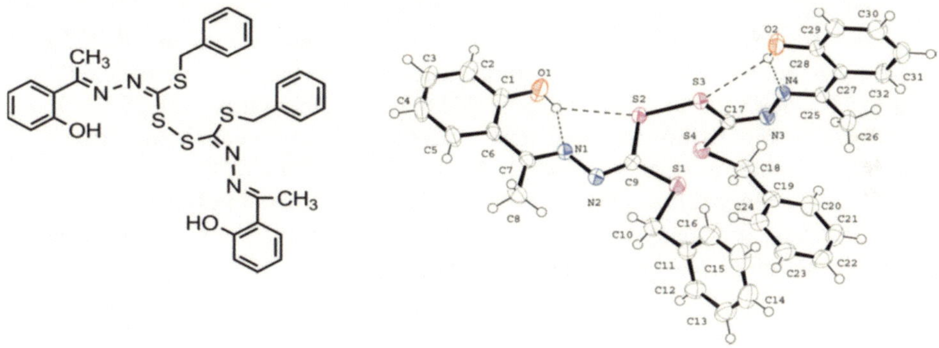

Figure 4.
The molecular structure of the 1,2-bis((benzylsulfanyl){2-[1-(2-hydroxyphenyl)ethylidene]hydrazin-1-ylidene}-methyl)disulfane.

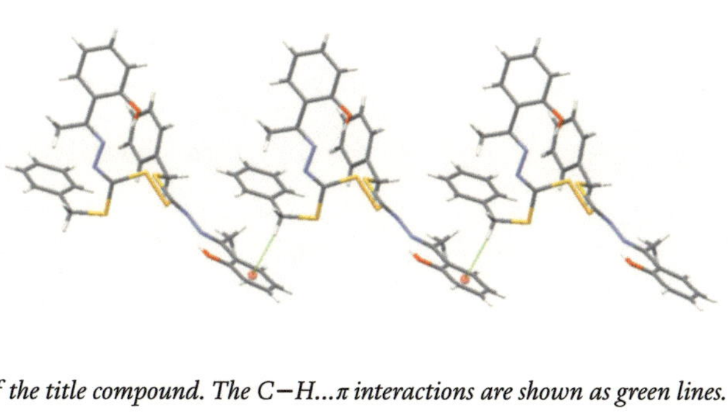

Figure 5.
A packing diagram of the title compound. The C—H…π interactions are shown as green lines.

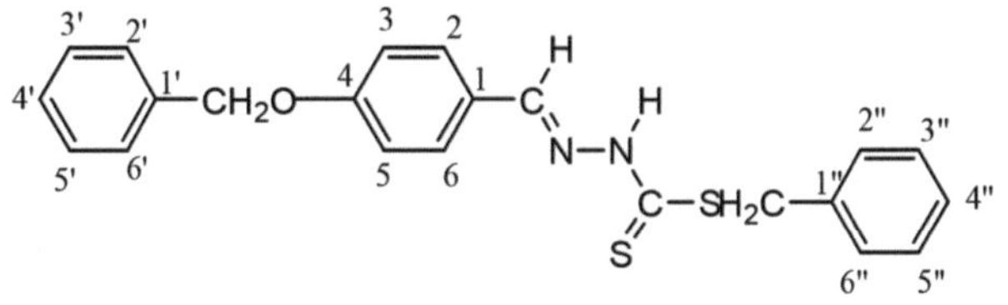

Figure 6.
S-Benzyl-β-N-(4-benzyloxybenzyl)methylenedithiocarbazate.

2.4.2.1 Synthesis of S-benzyl-β-N-(4-benzyloxyphenyl)methylenedithiocarbazate

4-Benzyloxybenzaldehyde (1.96 g, 10 mmol) in ethanol (25 mL) was added to a boiled solution of *S*-benzyldithiocarbazate (1.98 g, 10 mmol) in ethanol (40 mL), and the mixture was refluxed for 1h. The white solid that formed was separated by filtration and recrystallized from methanol as white crystalline solid (**Figure 7**). Yield: 3.74 g (95%). m.p. 140°C [25].

The ligand *S*-benzyl-*β*-*N*-(4-benzyloxyphenyl)methylenedithiocarbazate exists in thionetautomeric form both in the solid state and in solution. The Schiff base is bonded with the metal ions in a uni-negative bi-dentate mode, thereby forming four-coordinate inner complexes. Both the ligand and its complexes showed moderate antibacterial activity.

2.4.3 Synthesis of S-benzyl-β-N-(4-hydroxy-3-nitrophenyl)-methylenedithiocarbazate

Briefly, 4-hydroxy-3-nitrobenzaldehyde (1.67 g, 10 mmol) in acetonitrile (20 mL) was added in hot SBDTC (1.98 g, 10 mmol) in acetonitrile (40 mL) and refluxed for about 0.5 h. The resulting solution was cooled to room temperature (20–25°C). The orange-yellow solid that formed was separated, washed with ethanol, and dried in a vacuum over anhydrous $CaCl_2$—yield: 2.96g (81%), m.p. 171–172°C. The synthesized compound was recrystallized from a mixture of chloroform and petroleum ether (2:1; v/v) as large orange pyramid-shaped single crystals after 4 days of dissolution and slow evaporation at room temperature (**Figure 8**).

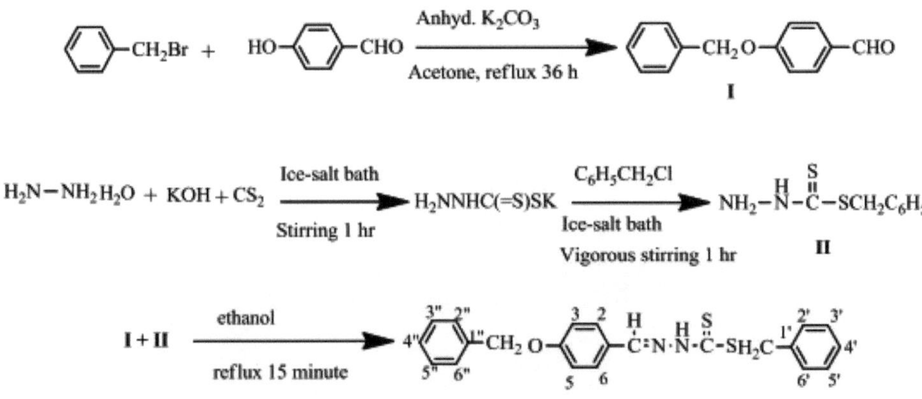

Figure 7.
Preparation of Schiff bases S-benzyl-β-N-(4-benzyloxyphenyl)methylenedithiocarbazate.

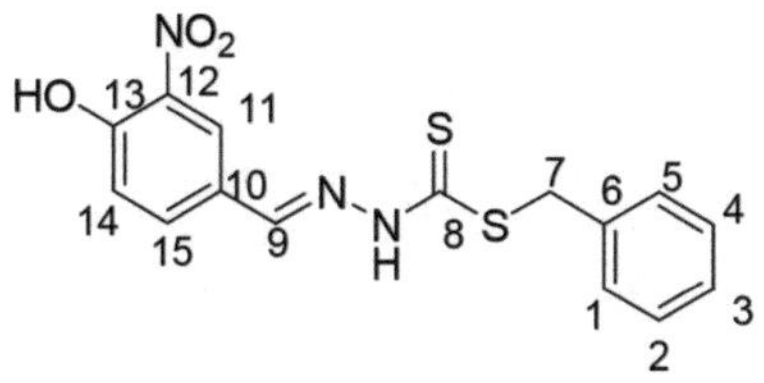

Figure 8.
Schiff bases S-benzyl-β-N-(4-hydroxy-3-nitrophenyl)-methylenedithiocarbazate.

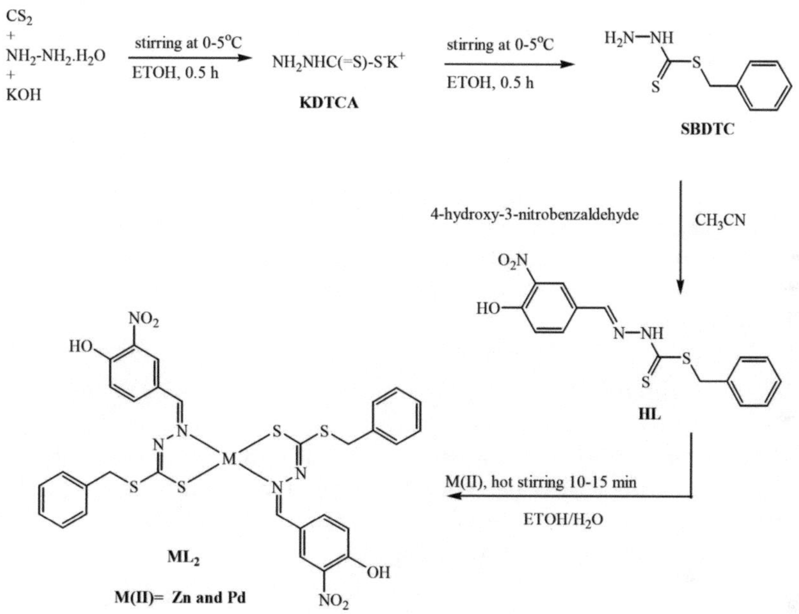

Figure 9.
Synthesis of Schiff base (HL) and its metal complexes.

The synthesized ligand was characterized using conventional techniques like nuclear magnetic resonance (NMR), UV-visible, infrared spectroscopy (IR), mass spectroscopic techniques, magnetic susceptibility measurement, and molar conductance. A single crystal X-ray crystallography data approved the proposed crystal structure of the ligand. Both in the solid and solution phase, the ligand continues to exist in its thione tautomeric form (**Figure 9**) [26].

3. Structural aspects and properties of Schiff bases

3.1 Molecular structure and bonding in Schiff bases

The molecular structure and bonding in Schiff bases are fundamental in their properties and reactivity [27]. Schiff bases are a class of organic compounds with a specific functional group, commonly represented as $R_1R_2C{=}NR_3$, where R_1, R_2, and R_3 can be various organic groups. Let us delve into the key aspects of the molecular structure and bonding in Schiff bases:

3.1.1 Functional group

The central feature of a Schiff base is the azomethine or imine functional group (C=N). This double bond (π bond) results from the linkage between a carbon atom (C) and a nitrogen atom (N). The carbon atom is often part of an organic group (R_1 or R_2), while the nitrogen atom can be part of another organic group (R3).

3.1.2 Geometry

The C=N double bond exhibits a planar geometry, with the carbon and nitrogen atoms lying in the same plane. This planar arrangement is a consequence of the π bond formed between C and N.

3.1.3 Bonding

The bonding in Schiff bases involves both sigma (σ) and pi (π) bonds. The sigma (σ) bond forms between the carbon atom (C) and one of the nitrogen atom's lone pairs, creating a single bond (C−N) and sp^2 hybridization around the carbon atom. In contrast, the combination of carbon and nitrogen atoms in the C=N double bond create the pi (π) bond.

3.1.4 Resonance structure

Schiff bases can exhibit resonance structures due to the presence of the π bond. The electrons in the π bond can delocalize, leading to resonance forms. This delocalization results in a partial double bond character in the C=N bond. The resonance forms contribute to the stability of the molecule.

3.1.5 Tautomeric forms

Schiff bases can exist in tautomeric forms. The most common tautomeric form is the imine tautomer, which has the C=N double bond. In some cases, Schiff bases can undergo tautomerization to form enamine tautomers.

3.1.6 Electron density and reactivity

The C=N double bond is polarized, with the carbon atom bearing a partial positive charge (δ+) and the nitrogen atom having a partial negative charge (δ−). This polarization makes the carbon atom electrophilic and the nitrogen atom nucleophilic, which influences the reactivity of Schiff bases. The electrophilic carbon can undergo additional reactions with nucleophiles, such as nucleophilic attack by various reagents.

3.1.7 Stereochemistry

The stereochemistry around the C=N bond can be important in certain reactions, particularly in the context of asymmetric synthesis. The orientation of substituents around the C=N bond can lead to different stereoisomers.

3.2 Spectroscopic characterization (UV-vis, IR, NMR, etc.) and analytical techniques

Spectroscopic characterization is a fundamental aspect of understanding the structural features and properties of Schiff bases. Various techniques, including UV-vis spectroscopy, infrared spectroscopy (IR), and nuclear magnetic resonance (NMR), play crucial roles in unraveling the molecular intricacies of Schiff bases [28].

3.2.1 UV-vis spectroscopy

UV–vis spectroscopy provides valuable insights into the electronic transitions of Schiff bases. It allows researchers to analyze the absorption of ultraviolet and visible light, providing information about the conjugation and electronic configuration of the carbon-nitrogen double bond. UV-vis spectroscopy is particularly useful for studying the chromophoric properties of Schiff bases, aiding in their identification and quantification.

3.2.2 Infrared spectroscopy (IR)

IR spectroscopy is employed to examine the vibrational modes of molecules, offering details about functional groups and molecular structure. In the context of Schiff bases, IR spectroscopy is instrumental in identifying characteristic peaks associated with the C=N stretching vibration, providing a fingerprint for the presence of the imine group.

3.2.3 Nuclear magnetic resonance (NMR)

NMR spectroscopy is a powerful technique for elucidating the structural aspects of Schiff bases. Proton NMR (^{1}H NMR) and carbon NMR (^{13}C NMR) offer information about the connectivity of atoms within the molecule, allowing precise determination of substituent positions and aiding in the verification of the proposed structure.

3.2.4 Analytical techniques

Beyond spectroscopy, other analytical techniques contribute to the comprehensive characterization of Schiff bases. Mass spectrometry (MS) can confirm molecular weight and elucidate fragmentation patterns, while elemental analysis provides information about the elemental composition of the compound. These techniques collectively contribute to a thorough understanding of the composition and structure of Schiff bases.

In summary, the integration of UV-vis spectroscopy, IR spectroscopy, NMR spectroscopy, and various analytical techniques enables a multifaceted characterization of Schiff bases. This holistic approach is essential for establishing their structural features, verifying synthetic outcomes, and laying the foundation for exploring their diverse properties and applications.

3.3 Thermodynamic and kinetic properties of Schiff bases

The study of thermodynamic and kinetic properties of Schiff bases is crucial for understanding their stability, reactivity, and potential applications. Researchers have delved into comprehensive investigations to unveil the intricacies of these properties [29].

3.3.1 Thermodynamic properties

Thermodynamic studies provide insights into the energetics of Schiff base formation and stability. The investigation encompasses enthalpy changes, Gibbs free energy, and entropy alterations during the reaction processes. Understanding these thermodynamic parameters is essential for predicting the feasibility and spontaneity of Schiff base reactions.

3.3.2 Kinetic properties

Kinetic analyses focus on the rate of Schiff base formation or decomposition. Experimental techniques, such as reaction progress monitoring and determination of reaction rates under varying conditions, are employed. These studies unravel the reaction mechanisms, activation energies, and the influence of different factors on the kinetics of Schiff base reactions.

The comprehension of thermodynamic and kinetic properties holds significance in diverse applications. Whether applied in catalysis, materials science, or medicinal chemistry, a thorough understanding of these properties contributes to the design and optimization of Schiff bases for specific purposes.

4. Applications of Schiff bases

Schiff bases, versatile compounds derived from the condensation of a primary amine and an aldehyde or ketone, find a multitude of applications across various scientific disciplines. Their diverse properties make them valuable in catalysis, medicinal chemistry, coordination chemistry, materials science, photovoltaics, sensing, corrosion inhibition, and environmental and industrial applications.

4.1 Catalysis and asymmetric synthesis using Schiff base ligands

Schiff bases play a pivotal role in catalysis and asymmetric synthesis, leveraging their versatile structures to influence the stereochemistry of reactions [15, 30, 31]. As ligands, Schiff bases form coordination complexes with various metal ions, imparting unique catalytic properties. These applications are integral in the synthesis of chiral compounds and are crucial in the pharmaceutical and fine chemical industries. They can coordinate with metal ions to form catalytically active complexes.

Catalysis and asymmetric synthesis:

1. *Catalysis*: Schiff base ligands exhibit catalytic activity in various reactions, including oxidation, reduction, and cross-coupling reactions. The coordination of metal ions with Schiff bases enhances the catalytic efficiency and selectivity of these processes.

2. *Asymmetric synthesis*: Schiff base complexes are widely employed in asymmetric synthesis, where the control of stereochemistry is essential. The chiral environment created by the Schiff base ligand in coordination with a metal center allows for the enantioselective formation of chiral compounds.

4.2 Medicinal chemistry: schiff bases as drug intermediates and bioactive compounds

Schiff bases have gained significant importance in medicinal chemistry, serving as versatile compounds with applications as drug intermediates and bioactive molecules [7, 8, 32, 33]. The unique structural features of Schiff bases, combined with their diverse pharmacological activities, make them valuable candidates for drug development.

Medicinal chemistry applications:

1. *Drug intermediates*: Schiff bases often act as intermediates in the synthesis of pharmaceutical compounds. They provide a versatile platform for introducing specific functionalities and optimizing the pharmacokinetic properties of drugs.

2. *Bioactive compounds*: Schiff bases exhibit a wide range of biological activities, including antibacterial, antifungal, antiviral, anti-inflammatory, and anticancer properties. These bioactive characteristics make Schiff bases attractive for designing novel therapeutic agents.

4.3 Applications of Schiff bases in coordination chemistry and metal-organic frameworks (MOFs)

Schiff bases play a crucial role in coordination chemistry, forming coordination complexes with metal ions and contributing to the construction of metal-organic frameworks (MOFs). These applications have far-reaching implications in catalysis, sensing, and materials science [34, 35].

Coordination chemistry and MOFs applications:

1. *Catalysis*: Schiff bases form coordination complexes with various metal ions, acting as ligands that influence the catalytic activity of the metal center. These complexes are employed in a variety of catalytic reactions.

2. *Sensing and detection*: the unique properties of Schiff bases make them suitable for sensing applications. Coordination complexes involving Schiff bases can act as sensors, detecting specific analytes through changes in their electronic or optical properties.

3. *Materials science*: Schiff bases contribute to the construction of metal-organic frameworks (MOFs), porous materials with diverse applications in gas storage, separation, and catalysis. MOFs incorporating Schiff bases can be tailored for specific functionalities.

4.4 Applications of Schiff bases in polymer science and materials development

Schiff bases play a significant role in polymer science and materials development, contributing to the design of innovative materials with tailored properties. Their versatile structures and reactivity make them valuable building blocks in the synthesis of polymers for various applications [36].

1. *Functional polymers*: Schiff bases can be incorporated into polymer chains to impart specific functionalities. These functional polymers find applications in drug delivery, sensors, and other advanced materials.

2. *Responsive polymers*: Schiff bases can be designed to respond to external stimuli such as pH, temperature, or the presence of specific ions. These responsive polymers are valuable in smart materials and controlled-release systems.

3. *Conducting polymers*: Schiff bases can be utilized in the development of conducting polymers, which find applications in electronic devices, sensors, and flexible electronics.

4. *Coating materials*: Schiff bases contribute to the synthesis of coatings with anticorrosive properties. These coatings protect surfaces from degradation in harsh environments.

4.5 Applications of Schiff bases in dye-sensitized solar cells and photovoltaics

Schiff bases play a significant role in the development of dye-sensitized solar cells (DSSCs) and photovoltaic devices, contributing to the advancement of sustainable and efficient energy conversion technologies [37].

1. *Light harvesting*: Schiff base dyes can act as sensitizers in DSSCs, absorbing light across a broad spectrum of wavelengths. The absorbed photons generate excited states that facilitate electron injection into the semiconductor material.

2. *Electron transport*: the imine linkage in Schiff base dyes provides a pathway for efficient electron transport, contributing to the generation of a photocurrent in the solar cell.

3. *Tunable properties*: the chemical versatility of Schiff bases allows for the design of dyes with tunable properties, including absorption spectra and redox potentials, enabling optimization for specific solar cell architectures.

4. *Stability and durability*: Schiff base dyes, when appropriately designed, can enhance the stability and durability of DSSCs, ensuring prolonged and efficient performance under varying environmental conditions.

4.6 Applications of Schiff bases as fluorescent probes and sensors

Schiff bases act as fluorescent probes and sensors, detecting specific analytes. Their fluorescence properties make them valuable in bioimaging and sensing applications [38–41].

1. *Metal ion sensing*: Schiff bases can selectively coordinate with metal ions, leading to changes in their fluorescence properties. This property is harnessed for the sensitive detection of metal ions in various samples.

2. pH sensing: Schiff bases can undergo protonation or deprotonation in response to changes in pH, resulting in alterations in their fluorescence intensity or emission wavelength. This property is exploited for pH-sensing applications.

3. *Biological molecule detection*: Schiff bases can be designed to interact selectively with specific biomolecules such as amino acids, proteins, or nucleic acids. Their fluorescence changes serve as indicators of the presence of these biological entities.

4. *Environmental monitoring*: Schiff base-based fluorescent probes are utilized for environmental monitoring, detecting pollutants, and assessing the quality of water or air based on specific chemical interactions.

4.7 Application of Schiff bases in corrosion inhibition

Schiff bases have corrosion-inhibitive properties, protecting metals from corrosion. They form a protective layer on metal surfaces, preventing the degradation of materials [42–44].

1. *Formation of protective films*: Schiff bases, characterized by their versatile molecular structures, have the capability to chelate with metal ions present on the surface of metals. This chelation process leads to the formation of protective films, effectively preventing corrosive agents from interacting with the metal substrate.

2. *Versatility and tailoring for specific applications*: the diversity in Schiff base structures allows for tailoring their chemical properties, enabling researchers to design compounds with specific corrosion-inhibiting characteristics. This versatility is advantageous for addressing the varying corrosion challenges posed by different metal substrates and environmental conditions.

3. *Enhanced corrosion resistance*: Schiff bases contribute to enhanced corrosion resistance by acting as barriers that impede the penetration of corrosive elements. The formation of a stable and protective layer on the metal surface retards the corrosion process, thereby extending the lifespan of the metallic structure.

4. *Potential for green corrosion inhibition*: the application of Schiff bases aligns with the principles of green chemistry as researchers explore environmentally friendly synthesis routes and sustainable practices for corrosion inhibition. This makes Schiff bases not only effective but also compatible with contemporary demands for eco-friendly solutions.

5. *Ongoing research and future prospects*: continuous research in the field of Schiff bases and corrosion inhibition aims to uncover new compounds with improved efficacy and application in diverse industries. The synergy of experimental studies and computational approaches contributes to a deeper understanding of the mechanisms involved, paving the way for innovative solutions and broader industrial applications.

In summary, the application of Schiff bases in corrosion inhibition stands as a promising area of research, offering tailored solutions for protecting metallic structures against the deleterious effects of corrosion. The versatility, effectiveness, and potential for environmentally friendly practices make Schiff bases valuable contributors to the field of corrosion science.

4.8 Application of Schiff bases in environmental and industrial contexts

Schiff bases find applications in various environmental and industrial contexts, including water treatment, catalysis for chemical transformations, and components of analytical techniques [45, 46].

1. *Metal ion chelation for wastewater treatment*: Schiff bases exhibit a remarkable ability to chelate metal ions. In environmental applications, this property is leveraged for wastewater treatment. By forming stable complexes with metal contaminants, Schiff bases aid in the removal of heavy metals from industrial effluents, contributing to the mitigation of environmental pollution.

2. *Catalysis*: acting as catalysts, Schiff bases enhance reaction efficiency in industrial processes, especially in pharmaceutical and petrochemical sectors.

3. *Corrosion inhibition*: Schiff bases protect metal surfaces, prolonging the life of industrial infrastructure.

4. *Solar cells*: their role as dye sensitizers contributes to the development of solar cells for sustainable energy.

5. *Antimicrobial coatings*: Schiff bases are employed to create coatings with antimicrobial properties on industrial surfaces.

6. *Sensor development*: used in sensor technology, Schiff bases help monitor environmental pollutants for quality control.

7. *Green chemistry initiatives*: Schiff bases support eco-friendly practices, aligning with green chemistry principles in industrial processes.

In essence, Schiff bases play a vital role in addressing environmental concerns and optimizing industrial processes for sustainability.

5. Recent advances in Schiff bases

5.1 Innovative and emerging applications

In recent years, Schiff bases have witnessed remarkable developments, leading to innovative and emerging applications across various scientific disciplines [16, 33, 47]. These advancements highlight the versatility and adaptability of Schiff bases in addressing contemporary challenges.

1. *Drug delivery systems*: Schiff bases have been explored in the development of advanced drug delivery systems, capitalizing on their ability to encapsulate and release therapeutic agents in a controlled manner.

2. *Nanomaterials and nanotechnology*: Schiff bases are integrated into nanomaterials for applications in nanotechnology, such as nanocatalysis, nanosensors, and drug delivery platforms.

3. *Biological imaging*: Schiff bases have gained prominence in biological imaging as fluorescent probes, enabling precise visualization of cellular processes and structures.

4. *Supramolecular chemistry*: recent advances in supramolecular chemistry involve the use of Schiff bases for constructing complex architectures, contributing to the development of functional materials.

5. *Photodynamic therapy*: Schiff bases have been explored in photodynamic therapy for cancer treatment, utilizing their photosensitive properties to induce cell death in targeted cancer cells.

5.2 Recent research and discoveries

In the dynamic field of Schiff bases, recent research and discoveries have unveiled novel applications, synthetic methodologies, and intriguing properties, contributing to the ongoing evolution of this versatile class of compounds [48–52].

1. *Catalytic applications*: recent studies have explored Schiff bases as catalysts in various reactions, showcasing their efficacy in asymmetric synthesis, cross-coupling reactions, and sustainable catalysis.

2. *Smart materials*: advances in Schiff bases have led to the development of smart materials that are responsive to external stimuli such as light, pH, or temperature, opening avenues for applications in sensing and responsive technologies.

3. Bioorthogonal chemistry: innovations in bioorthogonal chemistry involve the use of Schiff bases for selective labeling of biomolecules in living systems, providing new tools for bioimaging and diagnostics.

4. *Computational approaches*: recent discoveries leverage computational chemistry to design Schiff bases with tailored properties, accelerating the discovery of new compounds with specific functionalities.

6. Challenges and future directions

6.1 Current challenges and limitations in Schiff base research

Despite the versatility and widespread applications of Schiff bases, several challenges and limitations persist in the field. Addressing these issues is crucial for advancing Schiff base research and unlocking their full potential.

1. *Stability and reactivity*: Schiff bases, especially certain imine-based Schiff bases, may exhibit limited stability under certain conditions, impacting their long-term applications.

2. *Selectivity in metal complexation*: achieving high selectivity in metal complexation remains a challenge, particularly when dealing with multiple metal ions present in complex matrices.

3. *Biocompatibility*: while Schiff bases have promising applications in medicinal chemistry, issues related to biocompatibility and potential toxicity need to be thoroughly addressed for safe biomedical applications.

4. *Synthetic accessibility*: Some Schiff bases, especially those with intricate structures, may pose challenges in terms of synthetic accessibility, limiting their large-scale production and practical applications.

6.2 Future directions in Schiff base research

1. *Development of Stable Schiff Bases*: research efforts should focus on designing and synthesizing stable Schiff bases that maintain their integrity under various environmental and physiological conditions.

2. *Enhanced selectivity*: advancements in ligand design and coordination chemistry can contribute to the development of Schiff bases with improved selectivity for specific metal ions, reducing interference in complex samples.

3. *Biocompatible Schiff bases*: future research should aim to design and characterize Schiff bases with enhanced biocompatibility, ensuring their safe use in various biomedical applications.

4. *Green synthetic routes*: exploring sustainable and green synthetic routes for Schiff base synthesis can address concerns related to the environmental impact of traditional methods.

6.3 Future prospects and areas of potential growth in Schiff base chemistry

Schiff base chemistry holds promising avenues for future development, and researchers are actively exploring various areas to overcome current challenges and unlock new opportunities. The following outlines potential growth areas and future prospects in Schiff base research.

1. *Functional materials*: Schiff bases have shown potential in the design of functional materials for diverse applications, including sensors, catalysis, and drug delivery systems. Future research may focus on tailoring Schiff bases for specific functionalities and enhancing their performance in these applications.

2. *Supramolecular chemistry*: advances in supramolecular chemistry, involving the self-assembly of Schiff bases into well-defined structures, offer prospects for creating complex architectures with unique properties. This area holds potential for the development of new materials and devices.

3. *Biomedical applications*: the exploration of Schiff bases in biomedical applications, such as imaging agents, drug delivery systems, and therapeutics, represents a growing field. Future research may aim at improving biocompatibility, stability, and targeting capabilities for enhanced biomedical efficacy.

4. *Green chemistry*: green synthetic routes and environmentally friendly methodologies are gaining importance. Future prospects in Schiff base chemistry involve the development of sustainable synthesis methods, reducing environmental impact, and making the compounds more applicable in green chemistry initiatives.

6.4 The role of computational chemistry in advancing Schiff Base research

Computational chemistry plays a pivotal role in advancing Schiff base research by providing valuable insights into molecular structures, reaction mechanisms, and properties. The synergy between theoretical calculations and experimental studies enhances our understanding and guides the design of novel Schiff bases with tailored functionalities [52].

1. *Drug design and discovery*: computational methods aid in virtual screening and de novo drug design, accelerating the discovery of novel Schiff bases with therapeutic potential.

2. *Catalysis optimization*: computational chemistry contributes to the optimization of Schiff base-based catalysts, predicting reaction pathways and identifying key intermediates.

3. Electronic structure analysis: understanding the electronic structure of Schiff bases through computational methods aids in predicting their optical, redox, and magnetic properties.

4. *Biological interaction studies*: computational chemistry assists in studying the interaction of Schiff bases with biomolecules, providing insights into their potential applications in medicinal chemistry.

Computational chemistry, with its predictive power and ability to simulate complex molecular behaviors, continues to be instrumental in guiding and advancing Schiff base research across various scientific domains.

7. Expected contributions

This chapter will provide a comprehensive overview of Schiff bases, their synthesis, properties, and applications, with a focus on recent advancements. It will serve as a valuable resource for researchers, students, and professionals in the fields of chemistry, biology, and materials science. By presenting the novelties in Schiff bases, this chapter will contribute to the collective understanding of this versatile class of compounds.

8. Conclusion

In conclusion, this chapter comprehensively explores Schiff bases, revealing their synthesis, properties, and diverse applications across scientific domains. Key points include the historical context dating back to Hugo Schiff's discovery in the 19th century, examination of synthesis methods with a focus on green chemistry, elucidation of structural aspects, and exploration of versatile applications from catalysis to corrosion inhibition. Recent advances and future prospects, along with the emphasized versatility and importance of Schiff bases, underscore their crucial role in addressing contemporary challenges and shaping the landscape of modern chemistry.

Author details

Md. Hasibul Islam[1,2,3*] and Md. Abdul Hannan[4,5]

1 Rajshahi Sugar Mills Ltd. Bangladesh Sugar and Food Industries Corporation, Ministry of Industries, Bangladesh

2 Rajshahi University of Engineering and Technology (RUET), Rajshahi, Bangladesh

3 Polymer and Textile Research Lab, Department of Applied Chemistry and Chemical Engineering, University of Rajshahi, Rajshahi, Bangladesh

4 Zonal Poultry Farm Rajabarihat, Rajshahi, Ministry of Fisheries and Livestock, Bangladesh

5 Govt. Titumir College, Under University of Dhaka, Bangladesh

*Address all correspondence to: mhislamchem.bsfic@gmail.com

References

[1] Schiff H. Mittheilungen aus dem Universitätslaboratorium in Pisa: Eine neue Reihe organischer Basen. Justus Liebigs Annalen der Chemie. 1984;**131**:118-119. DOI: 10.1002/jlac.18641310113 (in German)

[2] Hernández-Molina R, Mederos A. Acyclic and macrocyclic Schiff Base ligands. In: McCleverty JA, Meyer TJ, editors. Comprehensive Coordination Chemistry II. Elsevier; Vol. 1. 2003. pp. 411-446. DOI: 10.1016/B0-08-043748-6/01070-7

[3] Schiff H. Eine neue Reihe organischer Diamine. Justus Liebigs Annalen der Chemie. 1866;**140**:92-137. DOI: 10.1002/jlac.18661400106

[4] Mannich C, Krösche W. Ueber ein Kondensationsprodukt aus Formaldehyd. Ammoniak und Antipyrin. Archiv Der Pharmazie. 1912;**250**(1):647-667. DOI: 10.1002/ardp.19122500151

[5] Yazdanbakhsh M, Takjoo R. Synthesis, crystal structure of new linear trinuclear isovalence Co (II)([Co3(H–L)2(L)2]), and visualizing intermolecular interactions with Hirshfeld surface method. Structural Chemistry. 2008;**19**:895. DOI: 10.1007/s11224-008-9356-1

[6] Núñez-Montenegro A, Carballo R, Vázquez-López EM. Synthesis and characterization of thiosemicarbazone derivatives of 2-ethoxy-3-methoxy-benzaldehyde and their rhenium(I) complexes. Polyhedron. 2008;**27**(13):2867-2876. DOI: 10.1016/j.poly.2008.06.018

[7] Babu KS, Daravath S, Swathi M, Ayodhya D, Shivaraj. Synthesis, anticancer, antibacterial, antifungal, DNA interactions, ADMET, molecular docking, and antioxidant evaluation of novel Schiff base and their Co(II), Ni(II) and Cu(II) complexes. Results in Chemistry. 2023;**6**:101121. DOI: 10.1016/j.rechem.2023.101121

[8] Aroua LM, Alhag SK, Al-Shuraym LA, Messaoudi S, Mahyoub JA, Alfaifi MY, et al. Synthesis and characterization of different complexes derived from Schiff base and evaluation as a potential anticancer, antimicrobial, and insecticide agent. Saudi Journal of Biological Sciences. 2023;**30**(3):103598. DOI: 10.1016/j.sjbs.2023.103598

[9] Zangrando E, Islam MT, Islam MAAAA, Sheikh MC, Tarafder MTH, Miyatake R, et al. Synthesis, characterization and bio-activity of nickel (II) and copper(II) complexes of a bidentate NS Schiff base of S-benzyl dithiocarbazate. Inorganica Chimica Acta. 2015;**427**:278. DOI: 10.1016/j.ica.2014.12.014

[10] Mohammed RM, Mahdi HA. Synthesis, characterization, antimicrobial, and antioxidant studies of new Schiff base ligands and their complexes with copper ion. History of Medicine. 2023;**9**(1):1065-1074. DOI: 10.17720/2409-5834.v9.1.2023.126

[11] Meena R, Meena P, Kumari A, Sharma N, Fahmi N. Schiff bases and their metal complexes: Synthesis, structural characteristics and applications. In: Akitsu T, editor. Schiff Base in Organic, Inorganic and Physical Chemistry. London, UK: IntechOpen; 2023. DOI: 10.5772/intechopen.108396

[12] Break MKB, Fung TY, Koh MZ, Ho WY, Tahir MIM, Elfar OA, et al. Synthesis, crystal structure, antibacterial

and In vitro anticancer activity of novel macroacyclic Schiff bases and their Cu (II) complexes derived from S-methyl and S-benzyl dithiocarbazate. Molecules. 2023;**28**(13):5009. DOI: 10.3390/molecules28135009

[13] Mumit MA, Islam MAAAA, Sheikh MC, Miyatake R, Mondal MOA, Alam MA. Synthesis, characterization and antimicrobial activity of a bidentate NS Schiff base containing S-allyl dithiocarbazate and its complexes. Journal of Molecular Structure. 2019;**1178**:583. DOI: 10.1016/j.molstruc.2018.10.046

[14] Hamid SJ, Salih T. Design, synthesis, and anti-inflammatory activity of some coumarin Schiff Base derivatives: In silico and in vitro study. Drug Design, Development and Therapy. 2022;**16**:2275. DOI: 10.2147/DDDT.S364746

[15] Singh A, Gogoi HP, Barman P, Guha AK. Novel thioether Schiff base transition metal complexes: Design, synthesis, characterization, molecular docking, computational, biological and catalytic studies. Applied Organometallic Chemistry. 2022;**36**:66-73. DOI: 10.1002/aoc.6673

[16] Khan S, Alhumaydhi FA, Ibrahim M, Al-Qahtani AS, Alshamrani M, Alruwaili AS, et al. Recent advances and therapeutic journey of Schiff Base complexes with selected metals (Pt, Pd, Ag, Au) as potent anticancer agents: A review. Anti-cancer Agents in Medicinal Chemistry. 2022;**22**(18):3086-3096. DOI: 10.2174/1871520622666220511125600

[17] Niu H, Huang Y, Bai X, Li X. Novel poly-Schiff bases containing 4,4′-diamino-triphenylamine as hole transport material for organic electronic device. Materials Letters. 2004;**58**(24):2979-2983. DOI: 10.1016/j.matlet.2004.05.007

[18] Zhao ZG, Kodaira T, Nagai N, Hakuta Y, Bando KK, Takashima H, et al. Self-standing microporous films of arrayed alumina nano-fibers including Schiff base molecules: Effect of the environment around the molecules on their photo-luminescence. Journal of Materials Chemistry. 2012;**22**(19):9738-9744. DOI: 10.1039/C2JM15798E

[19] Shukla M, Kulshrashtha H, Seth DS. Comparative study of the schiff bases by conventional and green method and antimicrobial activity. International Journal of Materials Sciences. 2017;**12**(1):71-76. ISSN 0973-4589

[20] Ali MA, Tarafdar MTH. Metal complexes of sulphur and nitrogen-containing ligands: Complexes of s-benzyldithiocarbazate and a schiff base formed by its condensation with pyridine-2-carboxaldehyde. Journal of Inorganic and Nuclear Chemistry. 1977;**39**:1785. DOI: 10.1016/0022-1902(77)80202-9

[21] Haque J, Srivastava V, Chauhan DS, Lgaz H, Quraishi MA. Microwave-induced synthesis of chitosan Schiff bases and their application as novel and green corrosion inhibitors: Experimental and theoretical approach. ACS Omega. 2018;**3**(5):5654-5668. DOI: 10.1021/acsomega.8b00455

[22] Nikpassand M, Fekri LZ, Sharafi S. An efficient and green synthesis of novel azo Schiff base and its complex under ultrasound irradiation. Oriental Journal of Chemistry. 2013;**29**(3):1041-1046. DOI: 10.13005/ojc/290326

[23] Anastas PT, Warner JC. Green Chemistry: Theory and Practice. Oxford: Oxford University Press; 2000

[24] Islam MAAAA, Sheikh MC, Islam MH, Miyatake R, Zangrando E. Crystal structure of 1,2-bis((benzylsulfanyl)

{2-[1-(2-hydroxyphenyl)ethyl-idene] hydrazin-1-ylidene}methyl)disulfane. Acta Crystallographica—Section E: Structure Reports Online. 2016;**72**:337. DOI: 10.1107/S2056989016002371

[25] Islam MH, Sheikh MC, Islam MAAAA. Studies on coordination chemistry and antibacterial activitiy of bidentate NS Schiff base derived from SBDTC and 4-benzyloxybenzaldehyde. Journal of Scientific Research. 2019;**11**(1):121-132. DOI: 10.3329/jsr.v11i1.37863

[26] Islam MH, Sheikh MC, Miyatake R, Zahan R, Islam MAAAA. Synthesis, characterization and bioactivity of bidentate NS Schiff base of S-benzyl dithiocarbazate (SBDTC) and its Zn(II) and Pd(II) complexes. Asian Journal of Chemistry. 2020;**32**(08):2091-2098. DOI: 10.14233/ajchem.2020.21870

[27] Smith MB, March J. March's Advanced Organic Chemistry. Reactions, Mechanisms, and Structure. 6th ed. Hoboken, New Jersey, Canada: John Wiley & Sons, Inc.; 2006. DOI: 10.1002/0470084960

[28] Silverstein RM, Webster FX, Kiemle DJ, Bryce DL. Spectrometric Identification of Organic Compounds. 8th ed. John Wiley & Sons; 2014. ISBN: 978-0-470-61637-6

[29] Ahmad N, Alam M, Wahab R, Ahmed M, Ahmad A. Synthesis, spectral and thermo-kinetics explorations of Schiff-base derived metal complexes. Open Chemistry. 2020;**18**(1):1304-1315. DOI: 10.1515/chem-2020-0168

[30] Meena DR, Aalam MJ, Chaudhary P, Yadav GD, Singh S. Synthesis and structural studies of Pd(II) complexes of bidentate Schiff bases and their catalytic activities as pre-catalysts in the Mizoroki-heck reaction. Polyhedron. 2022;**222**:115931. DOI: 10.1016/j.poly.2022.115931

[31] Kargar H, Fallah-Mehrjardi M, Behjatmanesh-Ardakani R, Munawar KS, Ashfaq M, Tahir MN. Synthesis, spectral characterization, SC-XRD, HSA, DFT and catalytic activity of a dioxidomolybdenum complex with aminosalicyl-hydrazone Schiff base ligand: An experimental and theoretical approach. Polyhedron. 2021;**208**:115428. DOI: 10.1016/j.poly.2021.115428

[32] Sohtun WP, Kathiravan A, Jhonsi MA, Aashique M, Bera S, Velusami M. Synthesis, crystal structure, BSA binding and antibacterial studies of Ni(II) complexes derived from dithiocarbazate based ligands. Inorganica Chimica Acta. 2022;**536**:120888. DOI: 10.1016/j.ica.2022.120888

[33] Iraji M, Salehi M, Malekshah RE, Khaleghian A, Shamsi F. Liposomal formulation of new arsenic Schiff base complex as drug delivery agent in the treatment of acute promyelocytic leukemia and quantum chemical and docking calculations. Journal of Drug Delivery Science and Technology. 2022;**75**:103600. DOI: 10.1016/j.jddst.2022.103600

[34] Alshaheri AA, Tahir MI, Rahman MB, Begum T, Saleh TA. Synthesis, characterisation and catalytic activity of dithiocarbazate Schiff base complexes in oxidation of cyclohexane. Journal of Molecular Liquids. 2017;**240**:486-496. DOI: 10.1016/j.molliq.2017.05.081

[35] Gupta KC, Sutar AK. Catalytic activities of Schiff base transition metal complexes. Coordination Chemistry Reviews. 2008;**252**(12-14):1420-1450. DOI: 10.1016/j.ccr.2007.09.005

[36] Nishat N, Rasool R, Parveen S, Khan SA. New antimicrobial agents: The synthesis of Schiff base polymers containing transition metals and their

characterization and applications. Journal of Applied Polymer Science. 2011;**122**(4):2756-2764. DOI: 10.1002/app.34100

[37] Abuamer KM, Maihub AA, El-Ajaily MM, Etorki AM, Abou-Krisha MM, Almagani MA. The role of aromatic Schiff bases in the dyes techniques. International Journal of Organic Chemistry. 2014;**4**:7-15. DOI: 10.4236/ijoc.2014.41002

[38] Bharali B, Talukdar H, Phukan P, Das DK. A new Schiff base based fluorescent sensor for Al (III) based on 2-hydroxyacetophenone and o-phenylenediamine. Journal of Fluorescence. 2020;**30**(4):751-757. DOI: 10.1007/s10895-020-02527-w J

[39] Krishnan U, Iyer SK. Iminothiophenol Schiff base-based fluorescent probe for dual detection of Hg2+ and Cr3+ ions and its application in real sample analysis. Journal of Photochemistry and Photobiology A: Chemistry. 2022;**425**:113663. DOI: 10.1016/j.jphotochem.2021.113663

[40] Aksuner N, Henden E, Yilmaz I, Cukurovali A. A highly sensitive and selective fluorescent sensor for the determination of copper (II) based on a schiff base. Dyes and Pigments. 2009;**83**(2):211-217. DOI: 10.1016/j.dyepig.2009.04.012

[41] Kumar J, Sarma MJ, Phukan P, Das DK. A new simple Schiff base fluorescence "on" sensor for Al^{3+} and its living cell imaging. Dalton Transactions. 2015;**44**(10):4576-4581. DOI: 10.1039/C4DT03932G

[42] Verma C, Quraishi MA, Alfantazi A, Rhee KY. Corrosion inhibition potential of chitosan based Schiff bases: Design, performance and applications. International Journal of Biological Macromolecules. 2021;**184**:135-143. DOI: 10.1016/j.ijbiomac.2021.06.049

[43] Kashyap S, Kumar S, Ramasamy K, Lim SM, Shah SAA, Om H, et al. Synthesis, biological evaluation and corrosion inhibition studies of transition metal complexes of Schiff base. Chemistry Central Journal. 2018;**12**(1):117. DOI: 10.1186/s13065-018-0487-1

[44] Arshad I, Qureshi K, Saleemi AS, Abdullah A, Bahajjaj AAA, Ali S, et al. Melamine–isatin tris Schiff base as an efficient corrosion inhibitor for mild steel in 0.5 molar hydrochloric acid solution: Weight loss, electrochemical and surface studies. RSC Advances. 2023;**13**(28):19301-19311. DOI: 10.1039/d3ra00357d

[45] Farag AA, Toghan A, Mostafa MS, Lan C, Ge G. Environmental remediation through catalytic inhibition of steel corrosion by Schiff's bases: Electrochemical and biological aspects. Catalysts. 2022;**12**(8):838. DOI: 10.3390/catal12080838

[46] Paul A, Silva TA, Soliman MM, Karačić J, Šljukić B, Alegria EC, et al. Benzimidazole Schiff base copper (II) complexes as catalysts for environmental and energy applications: VOC oxidation, oxygen reduction and water splitting reactions. International Journal of Hydrogen Energy. 2022;**47**(55):23175-23190. DOI: 10.1016/j.ijhydene.2022.04.271

[47] Jorge J, Del Pino Santos KF, Timóteo F, Vasconcelos RRP, Cáceres OIA, Granja IJA, et al. Recent advances on the antimicrobial activities of Schiff bases and their metal complexes: An updated overview. Current Medicinal Chemistry. 2023;**31**(17):2330-2344. DOI: 10.2174/0929867330666230224092830

[48] Abou-Omar MN, Attia MS, Afify HG, Amin MA, Boukherroub R,

Mohamed EH. Novel optical biosensor based on a nano-gold coated by Schiff base doped in sol/gel matrix for sensitive screening of oncomarker CA-125. ACS Omega. 2021;**6**(32):20812-20821. DOI: 10.1021/acsomega.1c01974

[49] Fatima B, Hussain D, Bashir S, Hussain HT, Aslam R, Nawaz R, et al. Catalase immobilized antimonene quantum dots used as an electrochemical biosensor for quantitative determination of H_2O_2 from CA-125 diagnosed ovarian cancer samples. Materials Science and Engineering: C. 2020;**117**:111296. DOI: 10.1016/j.msec.2020.111296

[50] Iye ÉN, Ribeiro MF, Katayama JM, Tadini MC, Balbino MA, Eleotério IC, et al. Electrochemical sensors containing Schiff bases and their transition metal complexes to detect analytes of forensic, pharmaceutical and environmental interest. A review. Critical Reviews in Analytical Chemistry. 2019;**49**(6):488-509. DOI: 10.1080/10408347.2018.1561242

[51] Chen H, Zhang Y, Xu C, Cao M, Dou H, Zhang X. Two π-conjugated covalent organic frameworks with long-term cyclability at high current density for lithium ion battery. Chemistry. A European Journal. 2019;**25**(68):15472-15476. DOI: 10.1002/chem.201903733

[52] Leelananda SP, Lindert S. Computational methods in drug discovery. Beilstein Journal of Organic Chemistry. 2016;**12**:2694-2718. DOI: 10.3762/bjoc.12.267

Chapter 2

Exploring Recent Advancements in Schiff Base Chemistry

Arif Ullah and Kareem Shah Bukhari

Abstract

Recent advancements in synthesizing Schiff bases, including sustainable and highly customizable approaches, are examined, offering insights into how their structures can be tailored for specific purposes. Schiff bases find applications in various fields, such as materials science, which contribute to developing polymers, metal-organic frameworks, nanocomposites, and medicinal chemistry, which serve as promising drug candidates. Additionally, they play crucial roles in coordination chemistry catalysis and have potential applications in emerging areas like nanotechnology and renewable energy, contributing to the development of nanomaterials and energy storage systems. This chapter provides a comprehensive overview of this synthesis and its applications, catering to researchers, students, and professionals looking to explore the multifaceted landscape of Schiff base compounds in contemporary organic chemistry.

Keywords: Green chemistry, pharmacological potential, complex molecules, condensation reactions, primary amines

1. Introduction

The scientific community recently became fascinated by an intriguing group of versatile organic compounds known as Schiff bases [1–3]. These compounds, which originate from the condensation of primary amines with carbonyl compounds, are distinguished by a dynamic interplay of chemical properties supporting their diverse applications. This chapter is an in-depth look at the complex area of Schiff Base Chemistry [4–6]. It attempts to give a thorough analysis of the synthesis techniques and a broad picture of the numerous applications that have sprung from the latest developments in this exciting field. It is essential to comprehend Schiff bases' historical roots to appreciate their contemporary relevance [7–10]. Schiff bases have evolved dramatically since Hugo Schiff's groundbreaking research in the late 1800s when he first described the condensation reaction between amines and aldehydes or ketones. Since their original discovery as synthetic intermediates, these substances have become adaptable building blocks with various functions. Revolutionary advancements in synthesizing Schiff bases have been made, indicating how organic chemistry constantly changes [10–12]. Green chemistry techniques, sonochemistry, and microwave-assisted synthesis are novel approaches that have enhanced conventional condensation reactions. Improving synthesis efficiency, these methods also correspond with the increasing focus on ecologically friendly and sustainable practices. These compounds' versatility is highlighted by the fact that Schiff base synthesis can now use a wide range of starting materials, including heterocyclic compounds and unconventional amines [13–15].

Schiff bases are distinguished by their structural diversity, evidence of the abundant possibilities inherent in their chemistry. Researchers have deciphered the complex architectures of Schiff base compounds using sophisticated spectroscopic techniques like mass spectrometry, X-ray crystallography, and NMR spectroscopy [16–19]. This structural realization makes understanding the relationship between molecular arrangement and properties possible, laying the groundwork for customizing Schiff bases for various uses. The emergence of Schiff bases as important participants in catalysis has greatly aided the development of effective and selective methods for organic synthesis. Schiff commands facilitate a spectrum of reactions by serving as ligands for transition metal catalysts, including oxidation, reduction, and cross-coupling [10, 20–22]. Changing the structure of the ligand allows for the modification of catalytic activity, which opens up new possibilities for synthesizing complex molecules with specific features. Schiff bases have emerged as an area of interest in medicinal chemistry due to their promising biological properties. Interest in using these substances for therapeutic purposes has increased due to their antimicrobial, antiviral, and anticancer characteristics. Structure-activity relationships are important in drug design because they have created hybrid compounds with improved pharmacological potential by combining Schiff bases with biologically active moieties [23–25].

The rich history, varied synthesis techniques, and plethora of applications of Schiff base chemistry become apparent as we progress through this chapter. There are many exciting developments and applications for Schiff bases, whose versatility inspires researchers to continue pushing the boundaries of chemical innovation. Join us as we explore the complexities of Schiff base chemistry and its revolutionary

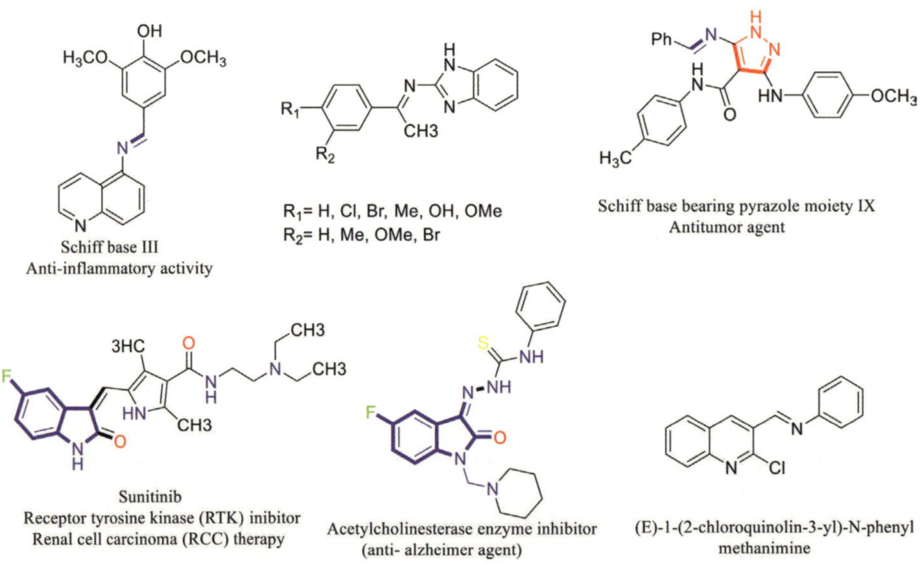

Figure 1.
Recently developed Schiff bases have found significant applications in therapeutics.

influence on modern organic chemistry, as scientists and enthusiasts are invited to participate (**Figure 1**).

2. Synthesis of unique Schiff bases and their applications

Ashraf S. Hassan, in 2023, a novel scientific finding [26], has demonstrated the possibility of a successful "one drug—multiple targets" strategy for treating complex medical conditions. 3-hydrazino-isatin was reacted with different aldehydes to form novel derivatives of isatin-based Schiff bases (2–7). Extensive *in vitro* evaluations encompassing antioxidant, antidiabetic, anti-Alzheimer, and anti-arthritic properties resulted from meticulous analyses that validated their structures. Compound 5a showed strong antioxidant activity, notable inhibition of α-amylase and acetylcholinesterase, and similar anti-arthritic properties to diclofenac sodium. Compound 5a's therapeutic potential was highlighted by *in silico* ADMET predictions and molecular docking simulations (**Figure 2**).

In 2022, Maryam Fawzi Habeeb created N-anthranilic acid [27] by starting the process with the Ullmann reaction, which involved combining derivatives of o-chlorobenzoic acid with o- and p-phenylenediamine. These compounds were cyclized using poly(phosphoric acid) (PPA) to produce products of amino acridone. The next step was to create terphthaldehyde mono Schiff bases by reacting terphthaldehyde with panisidine, o-, m-, or p-toluidine. These were then mixed with the derivatives of amino acridone to form additional Schiff bases. Different compounds were produced by the amino acridone derivatives reacting independently with terphthaldehyde, 4-(dimethyl amino) benzaldehyde, 2-formyl benzoic acid, and methyl-4-formyl benzoate. The structural characterization was done using FTIR spectroscopy, 13C NMR, and 1H NMR. The synthesized products demonstrated potential applications in medicinal chemistry and antibacterial research, as demonstrated by their antibacterial efficacy tests against *Klebsiella pneumoniae*, *Escherichia coli*, *Pseudomonas aeruginosa*, and *Staphylococcus aureus*. This thorough synthetic method yields a variety of compounds with various functions and possible biological applications (**Figure 3**).

Marijana Hranjec's study from 2023 explores [28, 29] the antiviral, antibacterial, and antiproliferative properties of novel Schiff bases made from benzimidazoles that have been N-substituted. The study examines the nature of substituents at the N atom of benzimidazole nuclei and their impact on the phenyl ring. The research uses *in vitro* testing to determine structure–activity relationships to assess the synthesized Schiff bases' antiviral efficacy against various antibacterial potentials, viruses against a range of bacterial strains, and antiproliferative effects on several human cancer cell lines. Although certain Schiff bases showed slight antiviral effects, these were noted at concentrations greater than those of reference medications. Precursor 23 showed broad efficacy against tested bacterial strains, while a subset of derivatives showed moderate antibacterial activity. Schiff base 40 stands out because it has a phenyl ring at the N atom and a 4-N, N-diethylamino-2-hydroxy substitution on the benzimidazole nuclei. IC50 ranges from 1.1 to 4.4 μM, indicating its potent antiproliferative action against multiple cancer cell lines. It was noted that Schiff base 40 had the strongest antitumoral effect against acute myeloid leukemia (HL-60). These results highlight the potential of the synthesized compounds, particularly Schiff base 40, as interesting subjects for additional investigation in creating cancer cell-targeting medicines (**Figure 4**).

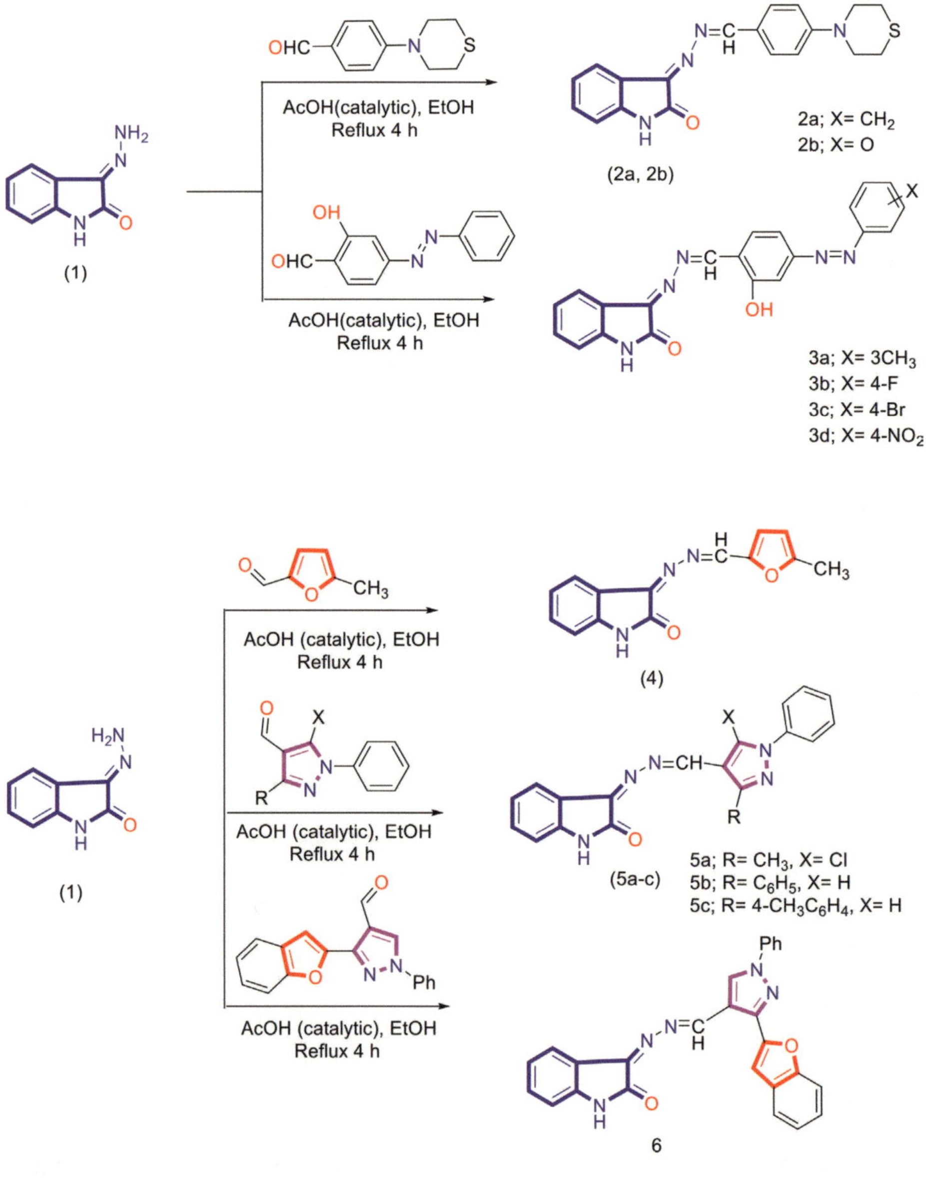

Figure 2.
Synthesis and production of Isatin-Centered azomethine and Isatin-Centered azomethine-arylazo compounds.

Nazish Shahab [29], successfully synthesized in three steps a series of novel Schiff's base hydrazone derivatives (1–24) based on polyhydroquinoline (PHQ) with good yields. Refluxing ethyl-2-(2-formylphenoxy)acetate, dimedone, ammonium acetate, and ethyl acetoacetate in ethanol for 6-7 hours produced the first compound (1). This was carried out in an unsymmetrical multi-component one-pot Hantzsch reaction. Compound 2 refluxed the resultant PHQ for 4–5 hours in absolute ethanol with hydrazine hydrate. The hydrazide was treated with various substituted aromatic/aliphatic aldehydes for the production of Schiff bases (3–24). Structural analysis was performed on all synthesized analogs using 1H, 13C, and LC-HRESI-MS spectroscopy. With IC50 values ranging from 5.26 to 25.17 μM, 12

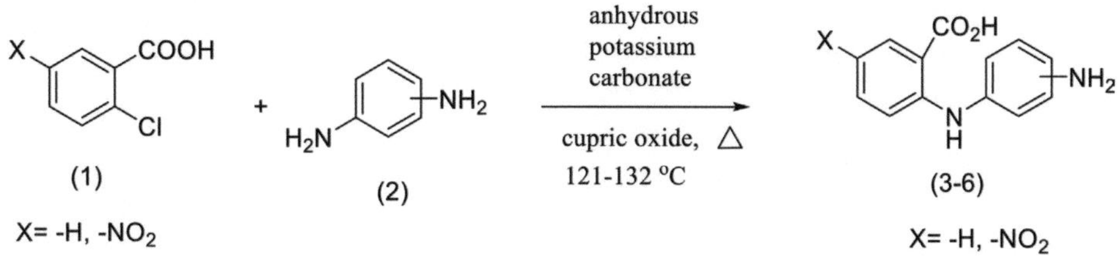

Scheme 1

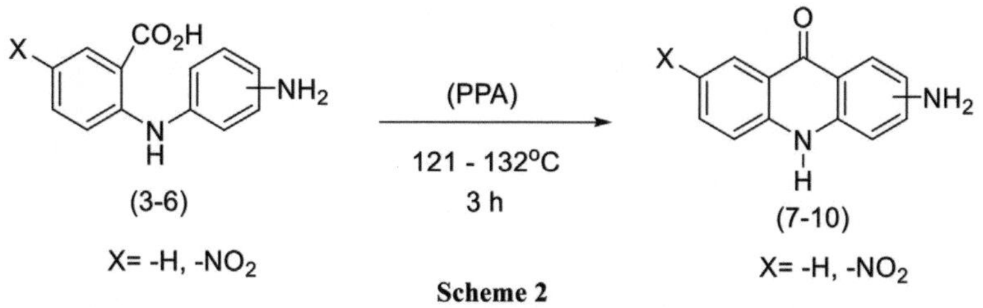

Scheme 2

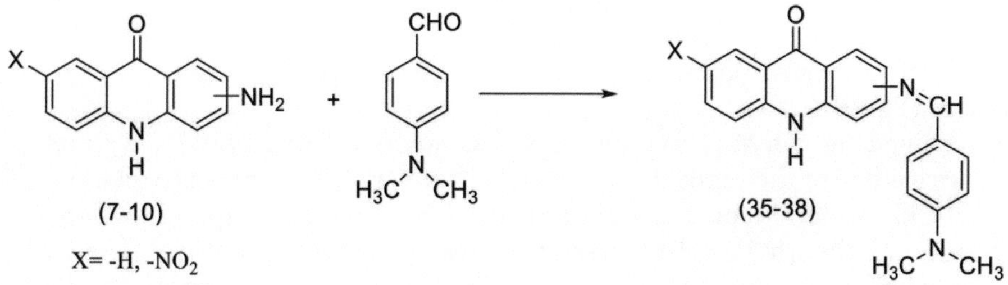

Figure 3.
Investigations into the creation, properties, and biological functions of new Schiff bases from amino acridone.

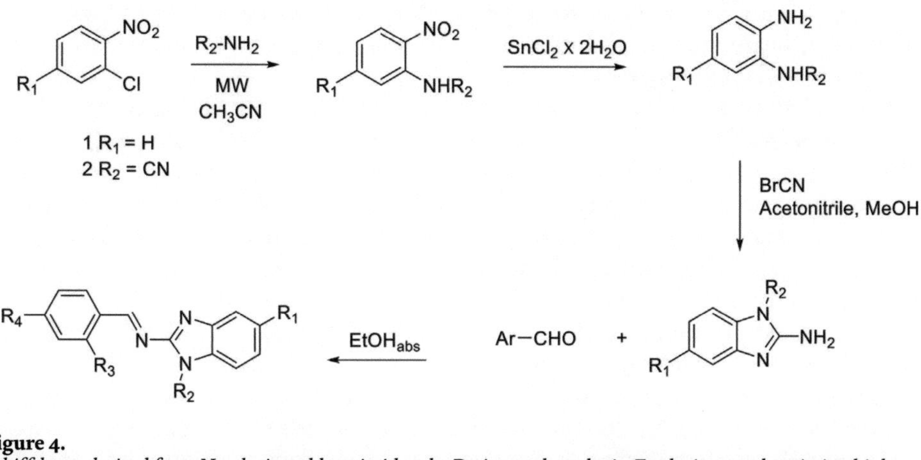

Figure 4.
Schiff bases derived from N-substituted benzimidazole: Design and synthesis: Exploring novel antimicrobial agents.

compounds (10, 14, 7, 9, 8, 17, 12, 19, 13, 15, 21, and 11) demonstrated significant activity against α-glucosidase. The move was higher than that of acarbose, which has an IC50 of 873.34 ± 1.67 μM, a standard α-glucosidase inhibitor. Molecular docking studies demonstrated that these active compounds efficiently occupied the

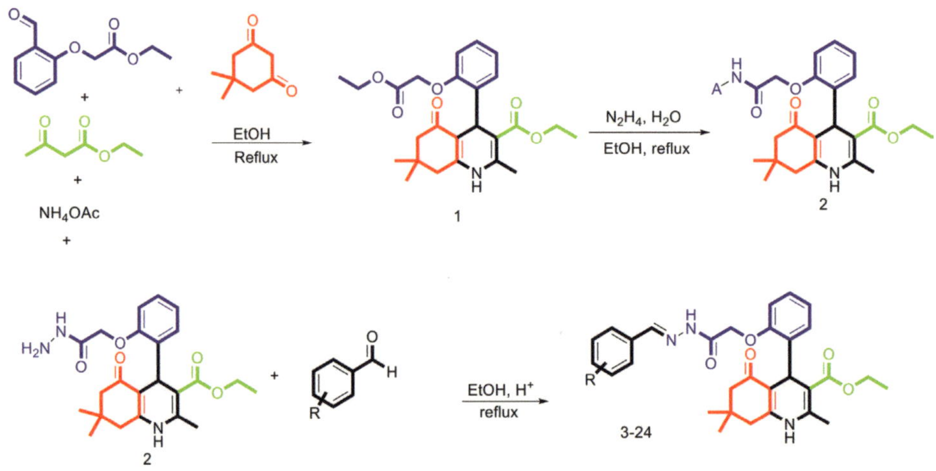

Figure 5.
Characterization and synthesis of novel hydrazone Schiff Base derivatives based on Polyhydroquinoline nucleus: Potential antioxidants and antibacterial agents.

α-glucosidase acarbose binding pocket, suggesting their potential as α-glucosidase inhibitors (**Figure 5**).

According to Najiah M. Alyamani, organic compounds' structural [1] symmetry or asymmetry is one of their most fundamental characteristics. Z-geometrical isomers are kinetically preferred, promiscuous structures with many biological functions. Beginning with the ethyl acrylate ester derivative (Z)-1 starting material, a new series of hybrid molecules containing 3,4,5-trimethoxybenzamide moieties and Schiff base were synthesized, and elemental microanalysis, 1H-NMR, and 13C-NMR spectroscopic studies were used to confirm their structural integrity. Colchicine was the reference material, and the MCF-7 breast cancer line was used as a test subject for the target hybrids' *in vitro* cytotoxic activity. Most recently developed combinations showed significant growth inhibition activity against the MCF-7 cells compared to the reference material (**Figure 6**).

As per Shahnazar Ali, in 2023, a novel class of thiazolidinones (TZD 5a-s) and azetidinones (AZT 6a-s) was developed, with their target being [2] the cox-2 enzyme (PDB protein 1PXX). 2-mercapto benzoxazole (1) was used as the starting material in conventional and microwave methods. The rat paw edema method was used to measure the anti-inflammatory activity *in vivo*, and DPPH free radical scavenging was used to evaluate the antioxidant activity *in vitro*. A comparative study compared microwave and conventional methods' yield, purity, and reaction time. Key intermediates, including ethyl 2-(benzoxazol-2-ylthio)acetate (2) and 2-(benzoxazol-2-ylthio)acetohydrazide (3), were formed during the synthesis process. Schiff bases (4a-s) were then created, and thiazolidinones (5a-s) and azetidinones (6a-s) were produced as a result (6a-s). The substances underwent tests for their ability to cause ulcers and reduce inflammation *in vivo* and their antioxidant capabilities *in vitro*. Three dimensions, robust CoMFA and CoMSIA models were obtained through QSAR analyses using SYBYL-X 2.1.1 software. The best models displayed acceptable prediction values and significant coefficients for a test set of eight compounds. After docking studies, molecules 5c, 5q, 5j, 6q, and 6 g showed acute ulcerogenic activity *in vivo* and favorable interaction energy with the COX-2 enzyme (protein 1PXX). These molecules were considered promising candidates. These findings provide valuable

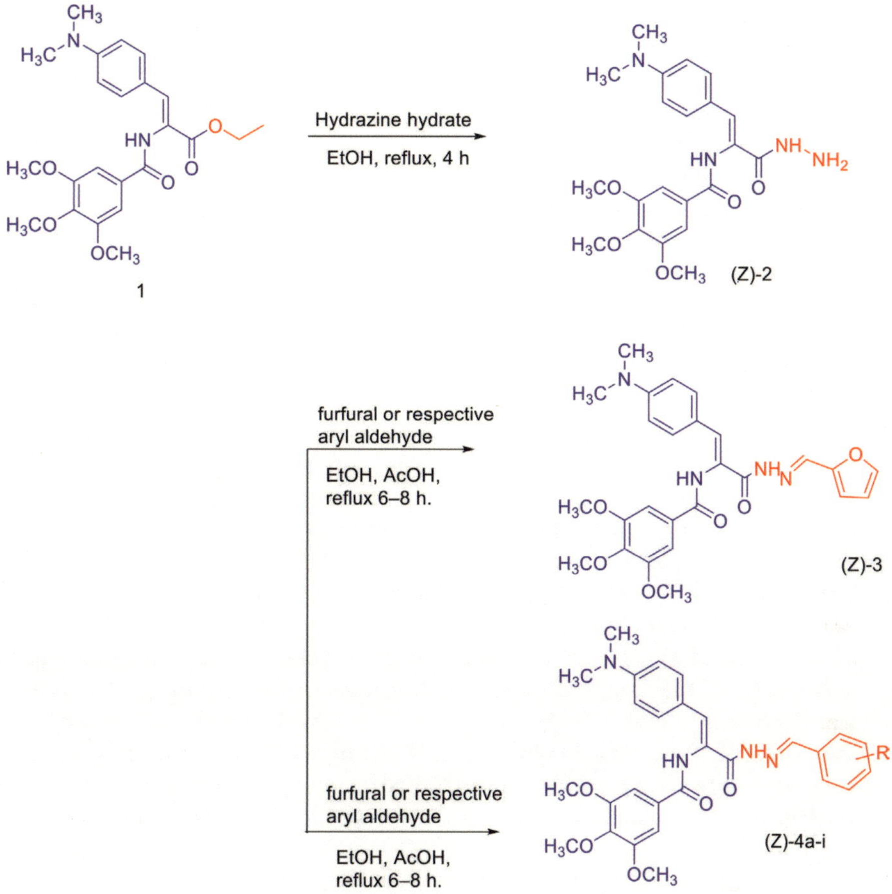

Figure 6.
Optimized synthetic routes for the design and elaboration of Schiff Base–TMB derivatives.

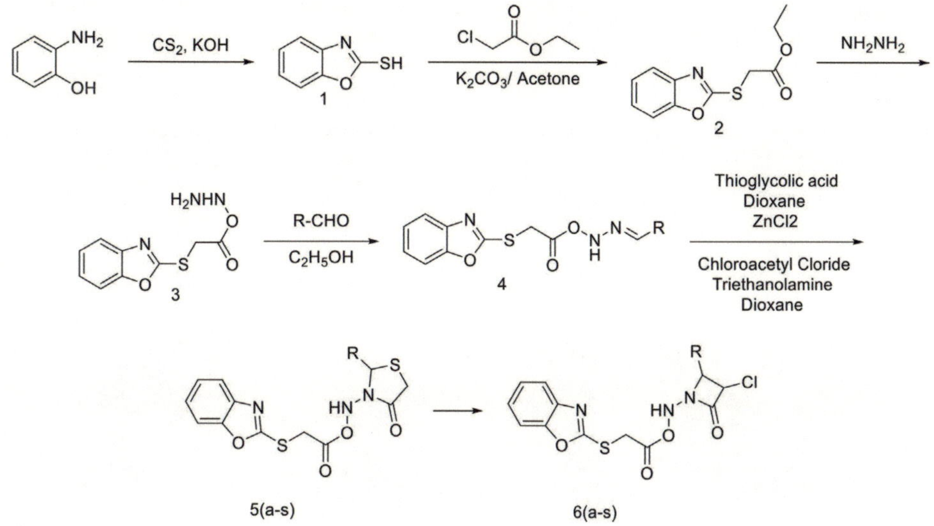

Figure 7.
Design and synthesis of benzoxazole-substituted Thiazolidinones and Azetidinones.

information that can be utilized to predict the compound activity and guide the synthesis of new molecules (**Figure 7**).

In February 2023, Kiran Singh and his team produced a novel [30] Schiff base known as five-methyl-3-((5-bromosalicylidene) amino)-pyrazole (HMBP) by refluxing 3-amino-5-methylpyrazole and 5-bromosalicylaldehyde in methanol at a 1:1 M constitution. Elemental analysis and spectral techniques (1 H NMR, IR, and mass) were used to characterize the HMBP. The coordination behavior of certain transition metal ions with newly synthesized Schiff base is also described. 1H NMR, fluorescence, elemental analyses, infrared spectra, electronic spectra, UV–vis, mass, and magnetic moments have all been used to characterize the complexes. The thermogravimetric assay (TG-DTG) has been used to conduct and explain the thermal study of the complexes. T Infrared spectra show that the Schiff base is a bidentate ligand linked to the metal ions by protonated phenolic O and imine N. The presence of water molecules is confirmed by thermal analysis. Quantum yields were measured in addition to the HMBP and all metal complexes' determined structures. According to a fluorescence study, metal complexes exhibit more intense fluorescence than the new Schiff base. To calculate the kinetics variables, the Coats-Redfern equation was applied. A variety of food-borne pathogens, such as Clostridium perfringens (MTCC 450), Escherichia coli (MTCC 433), Staphylococcus aureus sub aureus (MTCC 1144), Pseudomonas aeruginosa (MTCC 424), and fungi (Aspergillus niger, Aspergillus fumigatus, and Candida albicans), were tested against the synthesized ligand and its complexes. All of the complexes were found to have stronger biological activity toward individual organisms when compared to the new Schiff base (**Figure 8**).

In June 2022, Sheikh Abdul Majid's Schiff bases were [31] recognized as essential chemical scaffolds for applications in both industry and medicine. Because carbamate derivatives have so many uses, we report here the synthesis and characterization of a

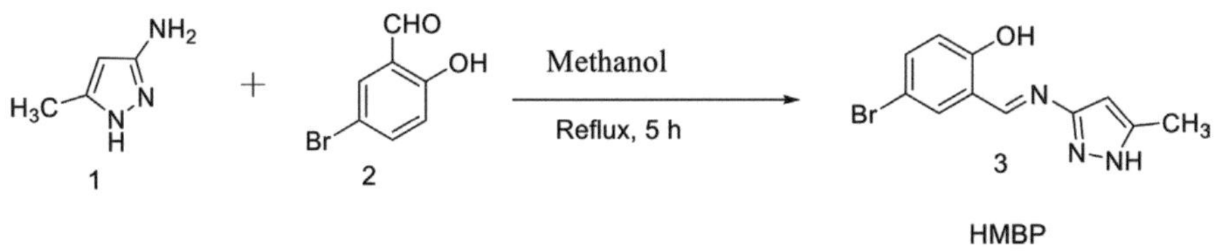

Figure 8.
Extensive studies on the synthesis of new Schiff Base 5-Methyl-3-((5-Bromosalicylidene) amino)-pyrazole and its transition metal complexes.

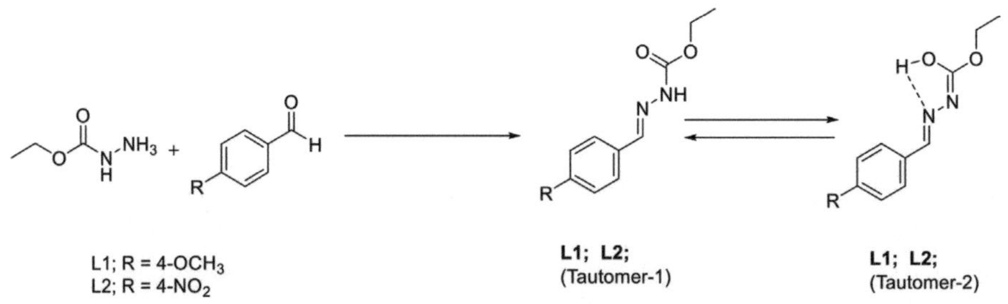

Figure 9.
Design and synthesis of Carbazate derivatives as innovative Schiff Base ligands for advanced coordination chemistry.

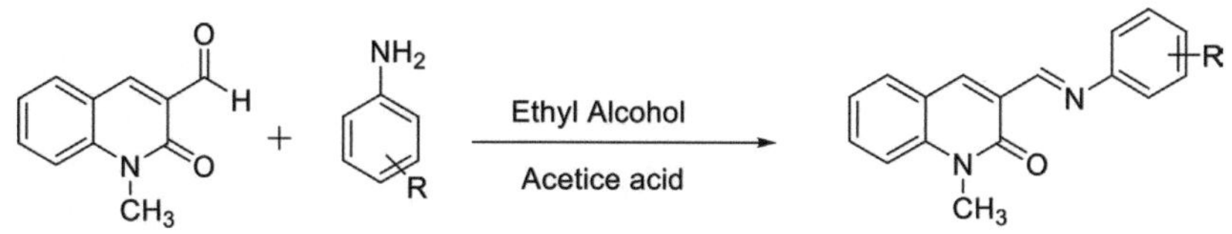

Figure 10.
Convenient synthesis, characterization, and biological evaluation of novel Schiff Base heterocycles.

new Schiff base ligand, (E)-ethyl 2-(4 methoxy benzylidene) hydrazine carboxylate, and 4-(nitrobenzaldehyde) methylcarbamate. The compound was characterized by combining B3LYP and 6-31 G(d,p)formalism with calculations from density functional theory and experimental data. The primary characterization techniques used for the elemental analysis of the mixture were FT-IR, UV-Vis, and NMR spectroscopic evaluations (**Figure 9**).

In 2023, Nandeshwarappa utilized condensation synthesis [32] to generate a sequence of (E)-1-methyl-3-((substituted phenylamino)methyl) quinoline-2(1H)-one Schiff bases (3a-j) containing the quinoline moiety (2a-j) from a solitary starting material, 1-methyl-2-oxo-1,2-dihydroquinoline-3-carbaldehyde (1). The structure of the recently synthesized compounds was verified by Fourier transformation infrared (FTIR), mass spectroscopy, elemental analysis, and proton (1H) and carbon (13C) nuclear magnetic resonance spectroscopy. The *in-vitro* antimicrobial activity of the synthesized compounds against gram-positive (Bacillus licheniformis and Bacillus cereus), gram-negative (Escherichia coli and Acinetobacter sp.), and antifungal (Aspergillus flavus and Pichnanomala) bacteria was tested using the agar well diffusion method. Furthermore, the compounds (3b, 3c, 3d, 3f, 3 g, and 3j) that show good activity are tested for *in-vitro* antitubercular activity against the mycobacterium tuberculosis H37Rv strain using the Micro-plate Almamar Blue Assay (MABA) method. Important information about the compounds' effectiveness against these strains is obtained from this process. The structural activities of the synthesized compounds were also discussed. The distinct substituent affixed to the phenyl ring in compounds 3a, 3c, 3d, 3f, and 3 g provides a strong argument (**Figure 10**).

3. Conclusion

In conclusion, investigating advances in Schiff base chemistry has shed light on the field's dynamic evolution and potential to have a revolutionary effect in modern chemistry. The experience has highlighted structural adaptability, creative synthetic approaches, and a wide range of uses for Schiff bases in fields such as medicinal chemistry and materials science. Schiff bases will significantly impact the direction of environmentally friendly and sustainable chemistry as we look toward the future. Their potential in cutting-edge fields like supramolecular chemistry, nanotechnology, and renewable energy sources offers intriguing chances for additional study and development. To help researchers and enthusiasts navigate this dynamic and exciting field of organic chemistry, the chapter has attempted to provide a comprehensive understanding of Schiff base chemistry.

Author details

Arif Ullah* and Kareem Shah Bukhari
School of Chemistry and Chemical Engineering, Shaanxi Normal University, Shaanxi, Xian, P.R. China

*Address all correspondence to: arifullah4star@gmail.com

References

[1] Alyamani NM. New schiff base–tmb hybrids: design, synthesis and antiproliferative investigation as potential anticancer agents. Symmetry. 2023;**15**:609

[2] Ali S, Omprakash P, Tengli AK, Mathew B, Basavaraj MV, Parkali P, et al. Synthesis of Novel Benzoxazole Derivatives: Exploration of Their Possible Pharmacological Profiles with Computational Studies. Polycyclic Aromatic Compounds. 2023;**43**:3853-3886

[3] Amirkhanov V, Rauf A, Hadda TB, Ovchynnikov V, Trush V, Saleem M, et al. Pharmacophores modeling in terms of prediction of theoretical physicochemical properties and verification by EXPERIMENTAL correlations of carbacylamidophosphates (CAPh) and sulfanylamidophosphates (SAPh) tested as new carbonic anhydrase inhibitors. Mini Reviews in Medicinal Chemistry. 2019;**19**:1015-1027

[4] Raczuk E, Dmochowska B, Samaszko-Fiertek J, Madaj J. Different Schiff bases—structure, importance and classification. Molecules. 2022;**27**:787

[5] Ullah A, Liu J, Khan AU, Khan QU, Guo F, Nazir S, et al. Diversification and design of novel aniline-pyrimidines via sonogashira/suzuki cross coupling reactions catalyzed by novel CLPN-Pd. ChemistrySelect. 2021;**6**:13551-13558

[6] Tidwell TT. Hugo (Ugo) Schiff, Schiff bases, and a century of β-lactam synthesis. Angewandte Chemie International Edition. 2008;**47**:1016-1020

[7] Qin W, Long S, Panunzio M, Biondi S. Schiff bases: A short survey on an evergreen chemistry tool. Molecules. 2013;**18**:12264-12289

[8] Jos S, Suja NR. Synthesis, characterization, and study of catalytic activity of chiral Cu(II) and Ni(II) salen complexes in the α-amino acid C-α alkylation reaction. Inorganica Chimica Acta. 2023;**547**:121323

[9] Şener N, Özkinali S, Altunoglu YC, Yerlikaya S, Gökçe H, Zurnaci M, et al. Antiproliferative properties and structural analysis of newly synthesized Schiff bases bearing pyrazole derivatives and molecular docking studies. Journal of Molecular Structure. 2021;**1241**:130520

[10] Wanwu D, Khan SU, Khan QU, Khan S, Alam A, Ullah A, et al. In-situ phase transformation and microstructure of reinforced aluminium matrix composites (aa-6061 t6+nicr np) with and without tic nanoparticles. European Journal of Materials Science and Engineering. 2022;**7**:28-36

[11] Shi L, Ge H-M, Tan S-H, Li H-Q, Song Y-C, Zhu H-L, et al. Synthesis and antimicrobial activities of Schiff bases derived from 5-chloro-salicylaldehyde. European Journal of Medicinal Chemistry. 2007;**42**:558-564

[12] Siddiqui HL, Iqbal A, Ahmad S, Weaver GW. Synthesis and spectroscopic studies of new Schiff bases. Molecules. 2006;**11**:206-211

[13] Murahashi S. Synthesis of phthalimidines from schiff bases and carbon monoxide. Journal of the American Chemical Society. 1955;**77**:6403-6404

[14] El-Sonbati A, Mahmoud W, Mohamed GG, Diab M, Morgan SM, Abbas S. Synthesis, characterization of Schiff base metal complexes and their biological investigation.

Applied Organometallic Chemistry. 2019;**33**:e5048

[15] Farooq M, Shujah S, Tahir K, Nazir S, Khan AU, Almarhoon ZM, et al. Ultra efficient 4-Nitrophenol reduction, dye degradation and Cr(VI) adsorption in the presence of phytochemical synthesized Ag/ZnO nanocomposite: A view towards sustainable chemistry. Inorganic Chemistry Communications. 2022;**136**:109189

[16] Kaya I, Vilayetoglu A, Mart H. The synthesis and properties of oligosalicylaldehyde and its Schiff base oligomers. Polymer. 2001;**42**:4859-4865

[17] Alemi AA, Shaabani B. Synthesis and characterization of a schiff base of p-tert-butylcalix[4]arene and its complex with copper(II). Acta Chimica Slovenica. 2000;**47**:363-369

[18] Uddin N, Rashid F, Ali S, Tirmizi SA, Ahmad I, Zaib S, et al. Synthesis, characterization, and anticancer activity of Schiff bases. Journal of Biomolecular Structure and Dynamics. 2020;**38**:3246-3259

[19] Liu J-J, Guo F-H, Cui F-J, Zhu J-H, Liu X-Y, Ullah A, et al. A biomass-derived N-doped porous carbon catalyst for the aerobic dehydrogenation of nitrogen heterocycles. New Journal of Chemistry. 2022;**46**:1791-1799

[20] Justin Dhanaraj C, Nair MS. Synthesis, characterization, and antimicrobial studies of some Schiff-base metal (II) complexes. Journal of Coordination Chemistry. 2009;**62**:4018-4028

[21] Imran S, Taha M, Ismail NH, Khan KM, Naz F, Hussain M, et al. Synthesis of novel bisindolylmethane Schiff bases and their antibacterial activity. Molecules. 2014;**19**:11722-11740

[22] Saleh EAM, Khan AU, Tahir K, Almehmadi SJ, Al-Abdulkarim HA, Alqarni S, et al. Phytoassisted synthesis and characterization of palladium nanoparticles (PdNPs); with enhanced antibacterial, antioxidant and hemolytic activities. Photodiagnosis and Photodynamic Therapy. 2021;**36**: 102542

[23] Jarrahpour A, Motamedifar M, Pakshir K, Hadi N, Zarei M. Synthesis of novel azo Schiff bases and their antibacterial and antifungal activities. Molecules. 2004;**9**:815-824

[24] Khan KM, Khan M, Ali M, Taha M, Rasheed S, Perveen S, et al. Synthesis of bis-Schiff bases of isatins and their antiglycation activity. Bioorganic & Medicinal Chemistry. 2009;**17**:7795-7801

[25] Elbakoush FE, Khan QU, Ullah M, Ullah A, Khan AU, Khan J, et al. Characterization and Thermal Degradation Study of Carbonization the Polyimide (PMDA/ODA)/Fe Composite Films. Transactions on Electrical and Electronic Materials. 2021;**22**:843-850

[26] Hassan AS, Morsy NM, Aboulthana WM, Ragab A. Exploring novel derivatives of isatin-based Schiff bases as multi-target agents: Design, synthesis, in vitro biological evaluation, and in silico ADMET analysis with molecular modeling simulations. RSC Advances. 2023;**13**:9281-9303

[27] Habeeb MF, Sheat AM. Studies on Synthesis, Characterization and Biological Activities of Novel Schiff Bases from amino acridone. Egyptian Journal of Chemistry. 2023;**66**:225-234

[28] Beč A, Cindrić M, Persoons L, Banjanac M, Radovanović V, Daelemans D, et al. Novel biologically active N-substituted benzimidazole

derived Schiff bases: Design, synthesis, and biological evaluation. Molecules. 2023;**28**:3720

[29] Shahab N, Kong D, Ali M, Alam A, Ur Rehman N, Ullah S, et al. Novel polyhydroquinoline Schiff's base derivatives: synthesis, characterization, in vitro α-glucosidase inhibitory, and molecular docking studies. Research on Chemical Intermediates. 2023;**49**:1-24

[30] Devi P, Singh K, Kubavat B. Synthesis, spectroscopic, quantum, thermal and kinetics, antibacterial and antifungal studies: Novel Schiff base 5-methyl-3-((5-bromosalicylidene) amino)- pyrazole and its transition metal complexes. Results in Chemistry. 2023;**5**:100813

[31] Majid SA, Mir JM, Bhat MA, Shalla AH, Pandey A, Hadda TB, et al. A pair of carbazate derivatives as novel Schiff base ligands: DFT and POM theory supported spectroscopic and biological evaluation. Journal of Biomolecular Structure and Dynamics. 2023;**41**:5499-5515

[32] Maruthesh H, Katagi M, Nandeshwarappa B. A convenient synthesis, characterization and biological evaluation of novel schiff base heterocycles as potential antimicrobial, antitubercular agents and their structural activity relationship. Current Chemistry Letters. 2023;**12**:759-768

Chapter 3

Design of Novel Functional Materials Using Reactions of Quinones with Aromatic Amines

Phuong-Truc T. Pham and Mamoun M. Bader

Abstract

This chapter presents examples for the successful use of Schiff base chemistry in the molecular design and synthesis of novel organic semiconductors composed of fused heterocyclic ladder oligomers as an alternative for oligoacenes. The materials resulting from reacting quinones with aryl amines (Schiff Bases) have proven to possess favorable properties for applications when compared with their oligoacene counterparts, including pentacene. Molecules obtained from reactions of 2,5-dichloro-1,4-benzoquinone and 2,3-dichloro-1,4-naphtoquinone with 2-aminophenols and 2-aminothiophenols discussed in this chapter. We also include one example of half a ladder molecule, which we usually use as a first step in the preparation of asymmetrical structures. Four molecules are discussed in this work. They are: benzo[5,6][1,4]thiazino[2,3-b]phenothiazine (**1**); 6,9-dichloro-5H-benzo[a]phenoxazin-5-one (**2**); benzo[a]benzo[5,6][1,4]thiazino [3,2-c]phenothiazine (**3**); and 13-nitrobenzo[a]benzo[5,6][1,4]oxazino[3,2-c]phenothiazine (**4**). Compounds **2** and **3** crystallize in the monoclinic space groups $P2_1/c$ with four molecules in the unit cell each, whereas compound **4** crystallizes in the orthorhombic Fdd2 space group with 16 molecules in the unit cell. Some of the most significant features in the single-crystal structures of three of these are also discussed. π–π Stack distances of 3.085 and 3.290 Å were also observed in compounds **3** and **4**.

Keywords: ladder oligomers, organic semiconductors, fused heterocyclic aromatic, π-stacking, crystalline organic semiconductors, dyes, anticancer drugs

1. Introduction

Conjugated organic molecules have attracted great interest due to their unique optical and electrical properties. They have potential for application devices such as field-effect transistors (FETs), light-emitting diodes (LEDs), and solar cells [1–4]. Among the most studied materials in this context were oligoacenes, such as pentacene, a molecule that has been heavily investigated for use as organic semiconductor [5]. Pentacene is rather expensive and is prone to oxidation with poor solubility in common organic solvents. Such undesired properties rendered pentacene and other acenes less attractive for practical applications. Several structural modifications of oligoacenes have been examined, including introduction of alkyl groups and electron-withdrawing groups to remedy some of the abovementioned undesired properties [6]. We would like to emphasize that both molecular structure and the way molecules arrange in the solid state (in single crystals and thin films) determine the properties of molecular materials and that device performance in both thin film and single-crystal configuration heavily depends on sloid state packing [4].

We and others have explored the modification of the carbon skeleton by replacing carbon atoms in pentacene for instance with heteroatoms, such as nitrogen, oxygen, and sulfur, by employing Schiff base chemistry to construct these heterocyclic aromatic compounds as alternatives to oligoacenes [3, 4]. Upon examining the structure of pentacene, one easily recognizes that heterocyclic compounds shown in **Figures 1** and **2** are reasonably accessible by employing Schiff base chemistry.

Phenothiazine, phenazine and phenoxazine have been used in a wide range of applications. These materials have also found a wide range of applications in medicinal chemistry and materials science [7–9]. Due to their planar rigid highly conjugated structures, these materials usually possess interesting optical properties, including fluorescence and nonlinear absorption. Many molecular structures can be prepared using Schiff base chemistry using appropriate quinones and aryl amines. The starting materials of choice to achieve this task in preparing *linear* pentacene-like heterocyclic molecules, such as the ones shown in **Figure 1**, for example, is to use the reaction of 2,5- dichloro-1,4-benzoquinone with 2-aminophenol; 1,2-phenylene diamine or 2-aminothiophenol. Starting with substituted aryl amines at various positions, one can modify the unsubstituted structures shown in (**Figures 1** and **2**) [7–9]. However, if one uses 2,3-dichloro-1,4-naphthoquinone instead of 2,5-dichloro-1,4-benzoquinone, *bent* or *Y-shaped* structures are obtained (**Figures 3** and **4**) [4, 10].

The double substitution when benzoquinones or naphthoquinones are used results in the extended linear and bent or Y-shaped structures shown above (**Figures 1** and **4**). These reactions have been well studied and many factors affecting yields and product distribution under various conditions of temperature, catalysts, and solvents were reported, and as such, that aspect will not be emphasized herein [7–9]. The focus of this chapter, however, is to showcase how Schiff base chemistry can be employed in preparing

Figure 1.
Example of the synthesis of symmetrical linear ladder oligomers.

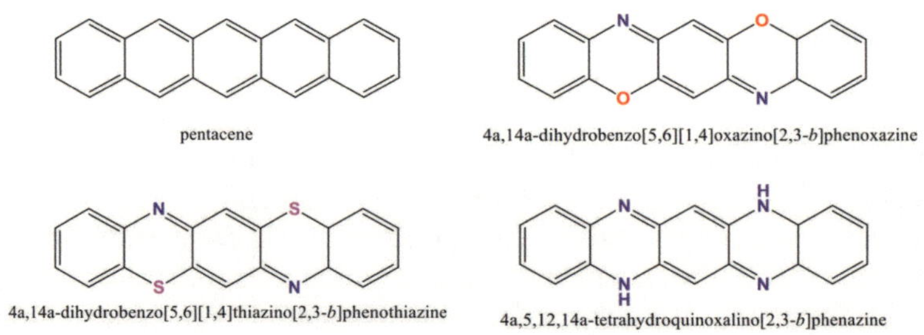

Figure 2.
Structural similarities between pentacene and symmetrical linear ladder oligomers readily accessible by reactions of aryl amines with 2,5-dichloro-1,4-benzoquinone.

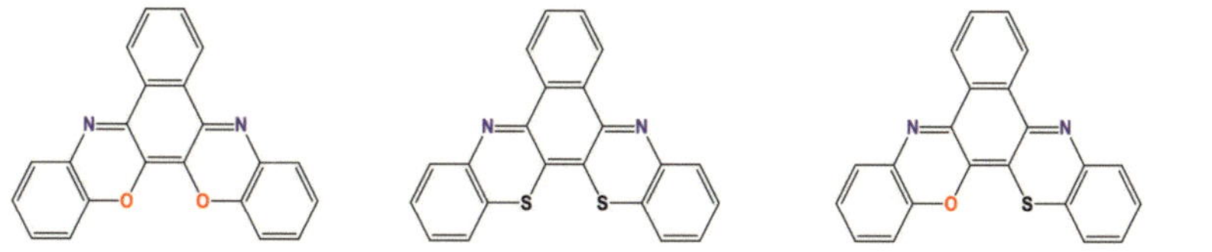

Figure 3.
Structures of symmetrical and asymmetrical bent or Y-shaped ladder oligomers readily accessible by reactions of aryl amines with 2,3-dichloro-1,4-naphthoquinone.

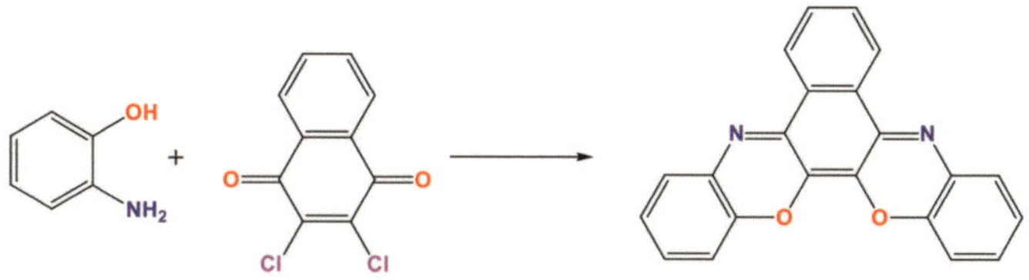

Figure 4.
Example of the synthesis of symmetrical bent or Y-shaped ladder oligomers.

and characterizing a few examples of ladder oligomers both linear and bent. The impact of shape and heteroatoms on solid state structures as revealed by single-crystal structural analysis and the observed *inter*- and *intra*molecular interactions will be discussed.

2. Results and discussion

All materials were prepared according to the published procedures [7–10]. In the choice of molecules to target, we attempted to address the impact of molecular geometry (linear vs. bent) and the presence of different heteroatoms on solid state structures of these materials and the types of inter- and intramolecular interactions in these structures. Surprisingly, despite this intensely researched area in terms of synthesis, biological activity, and optical properties, structural studies of these materials are scarce. In this chapter, we highlight our structural studies by means of single-crystal structural analysis on a group of known ladder-type oligomers. Four molecules, which were synthesized using Schiff base chemistry, are considered in this study. They are: benzo[5,6][1,4]thiazino[2,3-b] phenothiazine (**1**); 6,9-dichloro-5H-benzo[a]phenoxazin-5-one (**2**); benzo[a]-benzo[5,6][1,4] thiazino[3,2-c]phenothiazine (**3**); and 2-nitrobenzo[a]benzo [5,6][1,4] oxazino[3,2-c]phenothiazine (**4**) (**Figure 5**).

We started with targeting compound **1**, which has a backbone similar to the prototype organic semiconducting material, pentacene. The crystal structure of this molecule is known and was reported [11]. It was reported to crystallize in the monoclinic space group P21/c, a = 10.472(2), b = 5.430(3), and c = 12.356(10)A, Z = 2 [11]. We did not find a deposited CIF file associated with this structure or publication in the Cambridge Structural Database. The molecule is reported to be quasi planar with the central benzene ring of the molecule assuming a quinonoid structure with unequal C-C bond lengths. The authors also reported that the molecules of compound **1** are arranged in the unit cell in alternate parallel layers, with the planes of neighboring

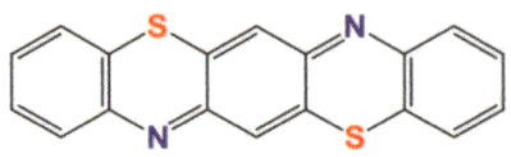

benzo[5,6][1,4]thiazino[2,3-*b*]phenothiazine
(1)

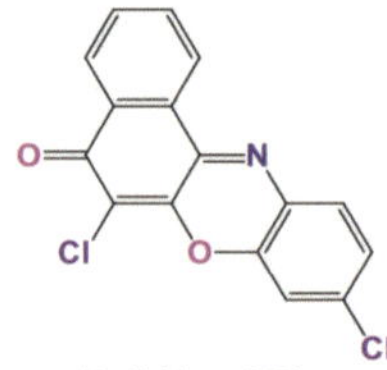

6,9-dichloro-5*H*-benzo
[*a*]phenoxazin-5-one
(2)

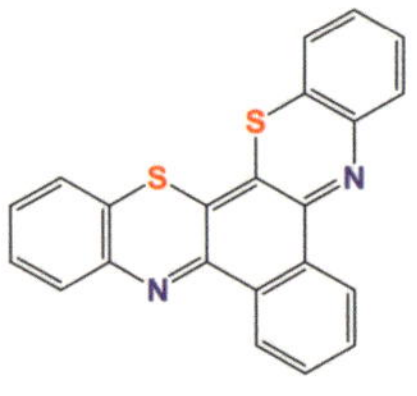

benzo[*a*]benzo[5,6][1,4]thiazino
[3,2-*c*]phenothiazine
(3)

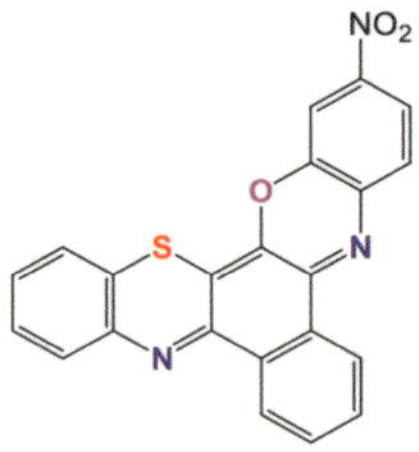

13-nitrobenzo[*a*]benzo[5,6][1,4]oxazino[3,2-*c*]phenothiazine
(4)

Figure 5.
*Structures of compounds **1-4**.*

layers almost perpendicular. Despite numerous efforts, we were unable to grow single crystals suitable for analysis. We then attempted to target the Y-shaped or bent (3-arm) ladder oligomer, compound **2** (the doubly substituted naphthoquinone); however, we always ended up with only half a ladder probably due to the reduced reactivity of the initial product as reported by Kafor [9]. We intend to use it for building other symmetric full ladders in future work. We, therefore, turned toward using the more reactive aminothiophenol and targeted the molecule benzo[5,6][1,4] thiazino[2,3-b]phenothiazine (**3**). This reaction worked smoothly and gave deep blue powder. This turned out to be an approach worthy of our efforts, as the molecule readily formed crystals both from solution (dichloromethane) and by sublimation (**Figure 6**). The calculated dipole moment for compound **3** is 0.65 Debye (DFT-B3LYP) compared with an essentially zero dipole moment for **1** [12]. We reported earlier on a single-crystal field-effect transistor using compound **3** [4].

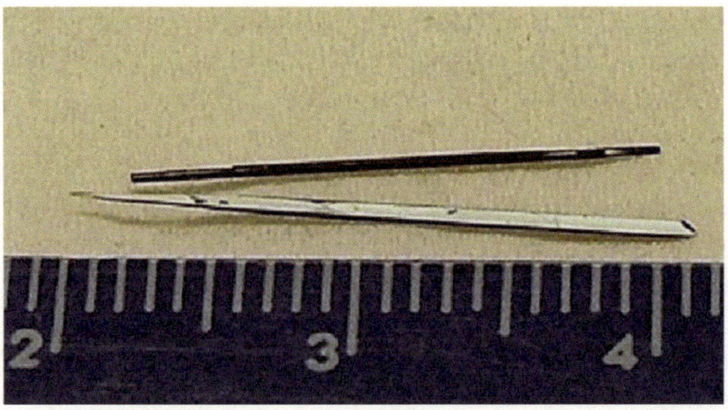

Figure 6.
*Crystals of compound **3** grown by physical vapor transport.*

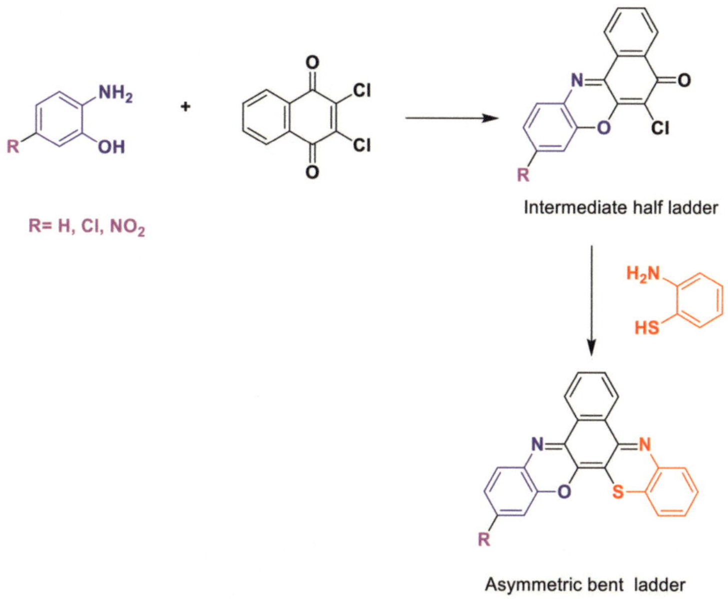

Figure 7.
Synthesis of asymmetric molecules (4, R = NO_2).

Molecule **4** was prepared in two steps, according to our published procedure, where half ladder was first prepared by reacting 2,3-dichloro-1,4-naphthoquinone with 4-nitro-2-aminophenol followed by reacting the resulting product with 2-aminothiophenol to access the asymmetrical molecule **4** (**Figure** 7) [10].

3. Solid state structures

Upon purification and crystallization from suitable solvent except for **3**, which was grown by sublimation, suitable crystals for X-ray diffraction were used. Compounds **2** and **3** crystallize in the monoclinic space groups $P2_1/c$ with 4 molecules in the unit cell each, whereas compound **4** crystallizes in the orthorhombic Fdd2 space group with 16 molecules in the unit cell. To the best of our knowledge, none of these structures were reported. The main common features of the three structures are that these molecules are nearly planar and tend to form π-stacks in the solid state. Both characteristics are favorable features for better device performance as they facilitate charge transfer.

Molecular structures and packing within the unit cells for compounds **2**, **3**, and **4** are shown in **Figure 8**. Compound **2** crystallizes in the monoclinic space group $P2_1/c$ with unit cell dimensions of: a = 16.4331(13) Å; b = 3.7613(3) Å; c = 20.6941(17) Å, and β = 98.5260(10)°; α = γ = 90° and 4 molecules in the unit cell. Compound **3** crystallizes in the monoclinic space group $P2_1/c$ with unit cell dimensions of: a = 11.7444(3) Å; b = 3.83000(10) Å; c = 35.8138(9) Å, and β = 97.5590(10)°; α = γ = 90° and 4 molecules in the unit cell. Compound **4** crystallizes in the orthorhombic space group Fdd2 with unit cell dimensions of: a = 6.7497(9) Å; b = 52.478(7) Å; c = 18.832(3) Å and α = β = γ = 90° and 16 molecules in the unit cell.

The dominant intermolecular interactions observed included: Cl...Cl close contacts of 3.729 Å and Cl...H 3.265 Å in compound **2**; C-H...S short distances of 2.942 and

3.221 Å in **3**; and N–O…H–C distance of 2.43 Å and C-H…N of 3.726 Å in **4**. π–π Stack distances of 3.085 and 3.290 Å were also observed in compounds **3** and **4**, respectively, while the optical band gap for **4** was estimated to be 1.85 eV (**Figures 8** and **9**) [10].

Finally, we have also carried out density functional theory (DFT) at the B3LYP/6-31G* level calculations for compounds 1-5 as gas phase isolated molecules, using

2

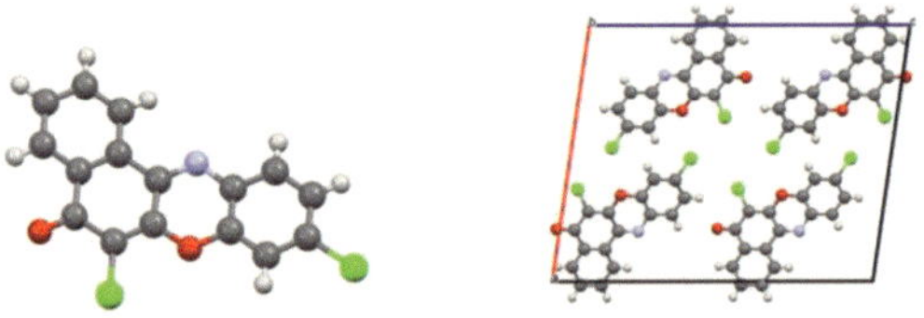

3

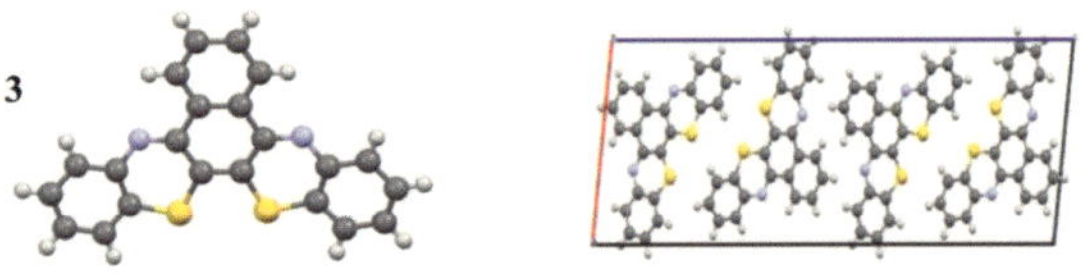

4

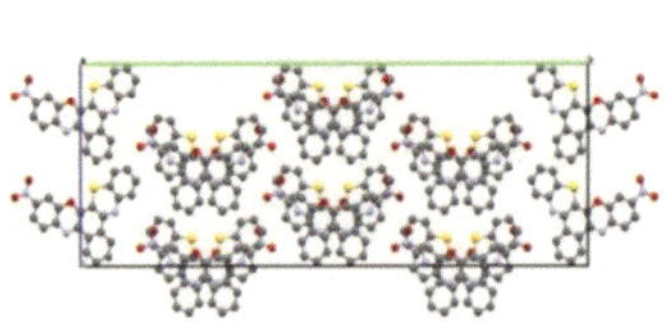

Figure 8.
Molecular structure and packing in the unit cell of 2, 3, and 4.

3

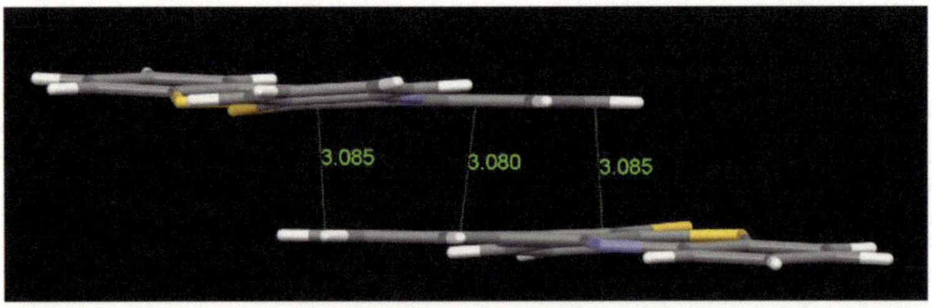

4

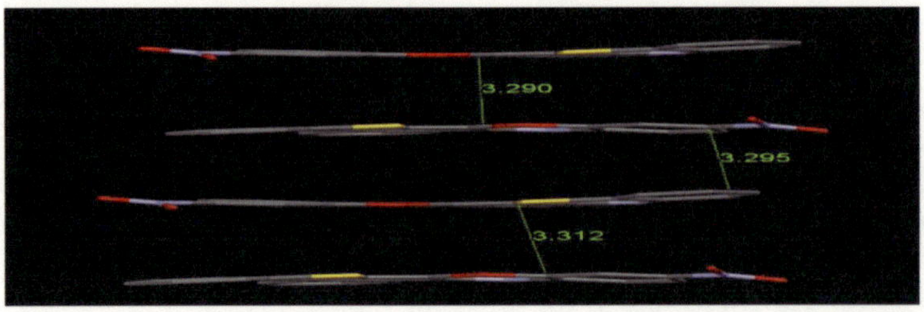

Figure 9.
π-π stacking in compounds 3 and 4.

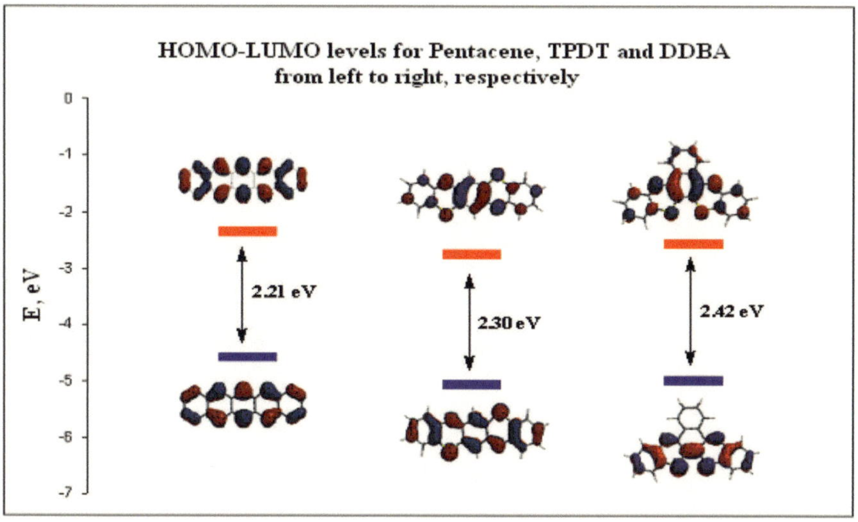

Figure 10.
*DFT calculated HOMO-LUMO levels (B3LYP) for compounds **1** (TPDT) and **3** (DDBA) compared with pentacene.*

Spartan 20 software [12]. The main features of these calculations are as follows: (i) Calculated molecular geometries were compatible with the experimentally determined ones. (ii) Linear ladder oligomer appears to have lower calculated band gaps compared to bent ones (**Figure 10**). It is worth noting that these calculations are carried out on isolated gas phase molecules.

4. Conclusions

In conclusion, we have demonstrated that Schiff base chemistry can be successfully applied to prepare interesting new materials suitable for various applications. The shape, (linear versus bent) of fused heterocycles containing N and S, plays a crucial role impacting the tendency of these materials to form crystals suitable for device fabrication. The presence of chlorine atoms results in Cl...Cl interactions, dominating the close intermolecular interactions within the unit cell. DFT calculations indicate that linear heteroacenes would have lower calculated band gaps. For example, compound **1** (linear) has a lower calculated band gap of 0.12 eV when compared with the bent molecule (**3**).

Acknowledgements

Funding from the Pennsylvania State University Research Development Grants and Alfaisal University IRG #18421 is acknowledged. The authors also acknowledge V. G. Young, Jr., and the X-Ray Crystallographic Laboratory, Department of Chemistry at the University of Minnesota, and C. Fiester, A, Bradely, and A. Nazzal of Wilkes University for help with synthesis of compound 4.

Author details

Phuong-Truc T. Pham[1*] and Mamoun M. Bader[2]

1 Department of Chemistry, Penn State Scranton, PA, USA

2 Department of Chemistry, Alfaisal University, Riyadh, Saudi Arabia

*Address all correspondence to: ptp2@psu.edu

References

[1] Buckley A, editor. Woodhead Publishing Series in Electronic and Optical Materials, Organic Light-Emitting Diodes (OLEDs). Woodhead Publishing; 2013. DOI: 10.1533/9780857098948. frontmatter. ISBN 9780857094254. Available from: https://www.sciencedirect.com/science/article/pii/B978085709425450021X

[2] Jacob MV. Organic semiconductors: Past, present and future. Electronics. 2014;**3**:594-597

[3] Mclean MR, Bader MM, Dalton LR, Devine RS, Steier WH. A Photophysical and structural study on dye-type organic molecules with potentially useful nonlinear optical properties. The Journal of Physical Chemistry. 1990;**94**:4386-4387

[4] Pham P-T, Yu X, Frisbie CD, Bader M. Single crystal field effect transistor of a Y-shaped ladder-type oligomer. Journal of Physical Chemistry C. 2008;**112**:7968-7971

[5] Laquindanum JG, Katz HE, Lovinger AJ, Dodabalapur A. Morphological origin of high mobility in Pentacene thin-film transistors. Chemistry of Materials. 1996;**8**:2542-2544. DOI: 10.1021/cm9603664

[6] Ye Q, Chi C. Recent highlights and perspectives on Acene based molecules and materials. Chemistry of Materials. 2014;**26**:4046-4056. DOI: 10.1021/cm501536p

[7] VanAllan JA, Reynolds GA. Polynuclear heterocycles. VI. The reaction of 2,3-dichloro-1,4-naphthoquinone with aromatic amines. The Journal of Organic Chemistry. 1963;**28**:1019-1022

[8] Agarwal NL, Schaefer W. Quinone chemistry. Reaction of 2,3-dichloro-1,4-naphthoquinone with o-aminophenols under various conditions. The Journal of Organic Chemistry. 1980;**45**:2155-2161

[9] Okafor CO. Synthesis of polycyclic mixed phenothiazine-phenoxazine organic dyes. Tetrahedron. 1988;**44**:1187-1194

[10] Fiester C, Pham PT, Bradley A, Nazzal A, Bader MM. Synthesis and characterization of a novel asymmetric fused ladder oligomer for applications as organic semiconductor. MRS Advances. 2023;**8**:889-893. DOI: 10.1557/s43580-023-00609-y

[11] Yukio I, Ichimin S, Michiko K, Naoki S, Hisao N. Molecular crystals and liquid crystals. Science and Technology, Section A: Molecular Crystals and Liquid Crystals. 1997;**296**:397-407

[12] Spartan 20. Wavefunction, Inc. Irvine, CA

Chapter 4

Significant Aspects of Heterocyclic Schiff Bases and Their Metal Complexes

Nabakrushna Behera, Tankadhar Behera, Jyotiprabha Rout and Sasmita Moharana

Abstract

Schiff bases are useful precursors for the synthesis of vital pharmaceutical and biochemical compounds due to their multifunctional transformations through different processes. Indeed, the heterocyclic Schiff bases have played a unique role in coordination chemistry owing to their wide-ranging potential bio-applications, such as anticancer, antibacterial, antifungal, etc. The exploration of novel heterocyclic Schiff bases and their metal complexes has certainly been progressing because of their proven usefulness as attractive key structures for the development not only in the field of biology but also in the field of catalysis, sensing, corrosion inhibitors, etc. The unusual characteristics of heterocyclic Schiff bases have resulted in many scopes, making significant advances from both the fundamental and applied perspectives.

Keywords: Schiff base, imine, anticancer, antimicrobial, metal complex, catalyst

1. Introduction

The discovery of Schiff base, more than one and a half century ago, has significantly stimulated chemical research, and has provided a crucial impetus for founding as well as rapidly expanding the domain of coordination chemistry that has persisted at a quick pace until now. Over the decades of intensive research, the Schiff base unit has been distinguished as a great versatile plinth for ligand design and complex synthesis in order to have their wonderful applications not only in the therapeutic but also in the catalysis and other fields. Schiff base ligands hold a privileged rank in the sphere of metal-ligand complex chemistry due to their characteristic advantage in synthesis. In fact, the Schiff base is a unique class of compounds that basically contain at least a moiety of imine (azomethine, –CH=N– and ketimine, >C=N–), generated by the condensation reactions between active carbonyl groups either from aldehyde or from ketone and primary amine compounds, wherein the nitrogen atom is bonded to an alkyl or aryl group (**Figure 1**) [1]. Thus, the common structural feature of Schiff bases is an imine group. Imine nitrogen is sp^2 hybridized and does contain a lone pair of electrons providing

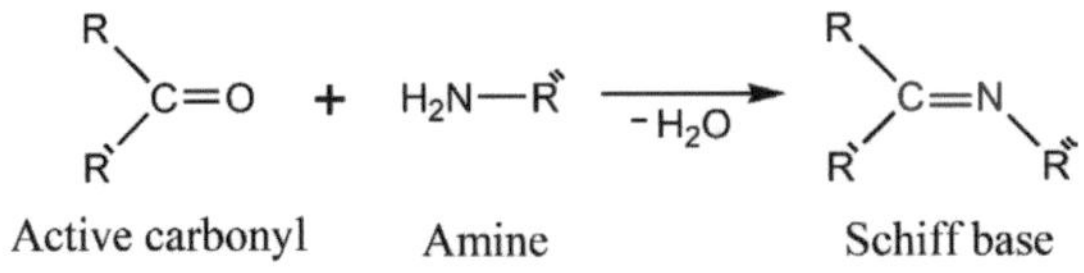

Figure 1.
A typical reaction scheme showing the generation of Schiff base.

considerable biological and chemical significances. The mechanism of Schiff base generation involves the attack of lone pair of electrons present on the nitrogen in primary amine (nucleophile) at the carbonyl carbon (electrophilic site) forming a tetrahedral intermediate. Subsequently, this process liberates water molecule and produces the imine group, irrespective of the methods or routes employed.

Schiff bases involving heterocyclic rings show several advantages due to the presence of more donor atoms, such as nitrogen (N), oxygen (O), and sulfur (S), and exhibit a vital role in coordination chemistry. However, among various heterocyclic molecules, nitrogen containing heterocycles have emerged as essential entities in the realm of medicinal chemistry. For instance, the benzimidazole and its derivatives represent groups of imperative chromophores with desirable medicinal properties and have dragged lots of attention in modern drug discovery [2–4]. As reported, the Schiff bases with benzimidazole moiety are extremely sensitive toward many human pathogens and microbial agents. Properties like antioxidant, antitumor, analgesic, anti-inflammatory, anticancer, and inhibitory are also their key functions [5]. Indole-derived Schiff bases have demonstrated extensive biological properties with anticancer, antitumor, and antimicrobial activities [6]. Such wide applications of heterocyclic Schiff bases (HSBs) are obviously rooted in the exclusive combination of the properties of the heterocyclic moiety, which exhibits a high degree of chemical stability but is amenable to several synthetic modifications.

One of the best abilities of almost all Schiff bases is to form strong complexes with metal ions. The formation of complex takes place owing to the coordination of donor atoms (*mostly* N, O, and S) through their unshared pair of electrons to the valence shell of metal ion, which in turn serves for alteration of electronic and steric surrounding of the metal center. Unlike reactive ligands, coordinated Schiff bases generally do not undergo irreversible transformations while modulating the reactivity of the metal ion at the core of the complex, and hence, these are called auxiliary ligands. Literature shows that Schiff base-derived metal complexes have played a promising role as drugs. The phenomenal pharmacological properties of HSBs have encouraged Schiff base complexation to a different level. HSB complexes have striking biological activities that include anticancer, antibacterial, antifungal, antimalarial, anti-inflammatory, antiproliferative, and antipyretic [7]. In addition, there are so many HSB complexes that exhibit excellent catalytic activity due to their exceptional stabilities. These versatile applications of HSBs and their metal complexes are undoubtedly attributed to their remarkable structural features, and some of the important applications have been discussed in this chapter.

2. Types of heterocyclic Schiff bases

The rewarding properties of Schiff bases are not restricted to a common integral feature of imine bond, rather they include the type of substituents present on its

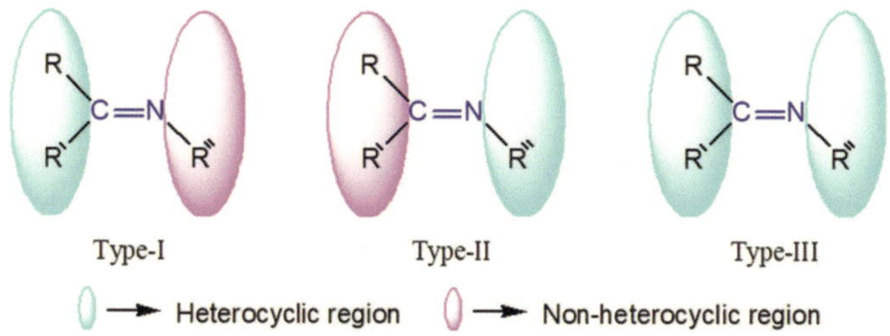

Figure 2.
Schematic representation of types of HSBs.

either side. The properties of Schiff bases can significantly be enhanced by modifying structurally with numerous scaffolds such as heterocyclic and substituted heterocyclic compounds on either or both sides of the imine unit. Although the conventional classification of Schiff base ligands is inherently dedicated to their denticities, here we have categorized HSBs into three types, such as Type-I, Type-II, and Type-III, depending on the presence of heterocyclic moiety(ies) around imine bond (**Figure 2**).

These HSBs could be obtained by different combinations of active carbonyls (aldehydes/ketones) with primary amines. For example, condensing heterocyclic carbonyl with non-heterocyclic amine or non-heterocyclic carbonyl with heterocyclic amine or heterocyclic carbonyl with heterocyclic amine offers Type-I, Type-II, and Type-III HSBs, respectively, or it could be achieved by the modifications of non-heterocyclic Schiff bases with the incorporation of heterocycle(s). The discussion pertaining to this chapter will purely be focused on the HSBs and related metal complexes containing at least one heterocyclic unit on either side of the imine group. Quite a few structural representations of HSBs (1–15) and some of their complexes (16–24) are shown in **Figure 3**.

3. Applications of heterocyclic Schiff bases and related complexes

Over the last decades, HSBs and their metal complexes find applications in versatile fields, such as biology, catalysis, and other industrial sectors. The goal of this chapter is to put forth a critical evaluation of these compounds taking into account current trends, the adaptability of Schiff bases with different heterocyclic moieties, and potential future research dimensions. Here, the applications of HSBs and their related metal complexes have been emphasized mostly on three crucial aspects, namely, anticancer activity, antimicrobial activity, and catalytic activity. Apart from this, a miscellaneous application section has been incorporated which basically provides a glimpse of other applicational diversities of HSBs or HSB-derived metal complexes.

3.1 Anticancer activity

Cancer is considered one of the lethal diseases after the cardiovascular disease. This is caused owing to the abnormal and uncontrolled cell growth. Continuous studies in the last few decades and development of pertinent drugs have raised hope to curb the cancer. It has been observed that indole containing alkaloids, such as vinblastine, vincristine, vinorelbine, and vindesine, are used as anticancer agents, due to their ability to interact with various receptors. However, several Schiff bases that

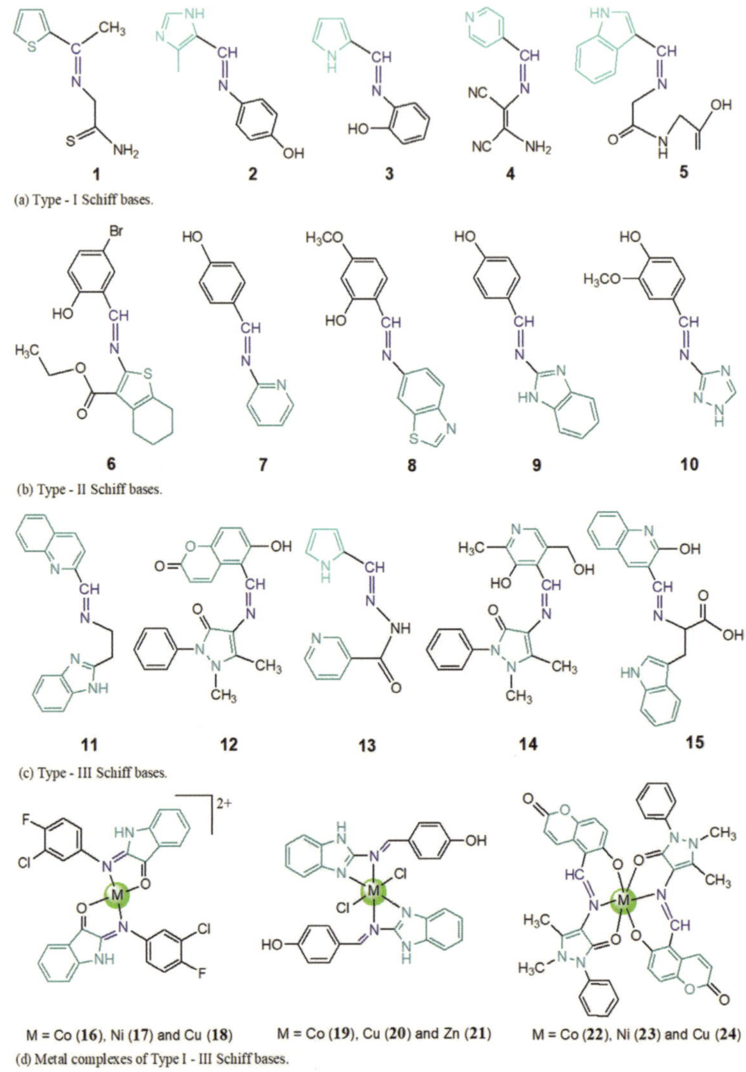

Figure 3.
Some examples of different types of HSBs 1–15 and complexes 16–24.

contain indole moiety(ies) are explored to possess the properties of anticancer against different types of cancer cells. Ghaidan et al. synthesized a series of HSBs (25–27) containing indole units (**Figure 4**), and these bases were found to be very effective for the inhibition of AMJ13 breast cancer cells while bringing no damage to normal cell growth [8]. Among these, dimethyl-substituted Schiff base 27 exhibits better activity even at its lesser dose of 10 μg/mL, inhibiting 60% after 72 h of exposure, and giving a scope for the researchers to reconsider it as an anticancer drug.

The ability of indole-based Schiff bases as active anticancer agents has been exploited by Trivedi et al. while they synthesized quite a few numbers of 3-substituted indole Schiff base derivatives consuming N-substituted indole-3-carboxaldehyde and substituted hydrazide (**Figure 5**) [9]. The anticancer activities of these compounds were tested against the A549 cell line (lung cancer) using MTT (3-[4,5-dimethylthiazol-2-yl]-2,5 diphenyl tetrazolium bromide) assay and showed

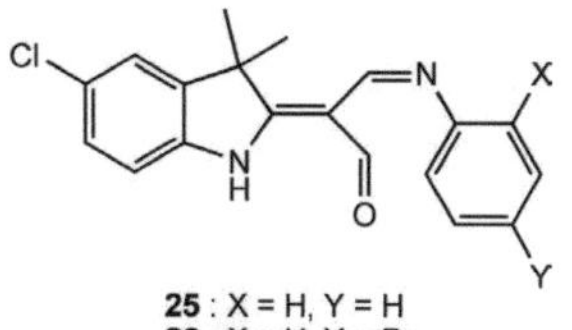

Figure 4.
Indole-derived HSBs.

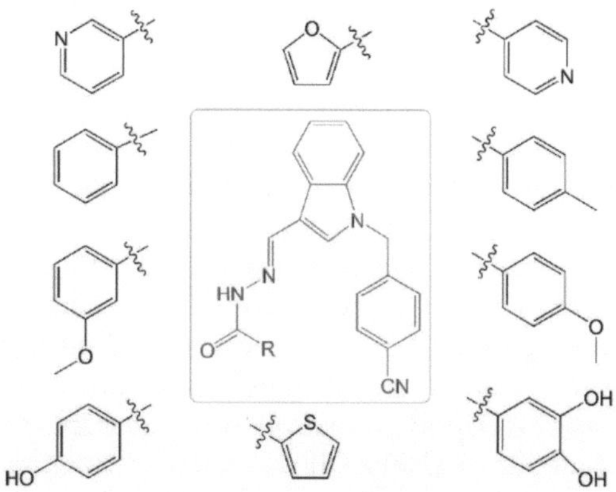

Figure 5.
Structure of indole-based Schiff base derivatives with various 'R' substituents.

moderate to excellent anticancer activity in comparison to the Osimertinib, a commercially available anticancer drug to treat lung cancer. Interestingly, the presence of furan ring in the Schiff base has produced better and remarkable results over other derivatives along with Osimertinib [9]. Thus, substituting suitable heterocycles for the R-group generates improved results than those obtained with non-heterocycles, signifying the importance of the heterocycles in the development of anticancer drugs.

Indole-3-carbaldehyde thiosemicarbazone-derived palladium complexes 28–34 (**Figure 6**) were synthesized by Haribabu et al. [10]. The anticancer activities of these metal complexes were evaluated against HepG2, A549, and MCF7 cell lines using MTT assay, and were found to have good activity against HepG2 cells. However, the complexes 31 and 32 exhibit potential cell killing. Al-Shemary et al. have found the possible anticancer activities of triazole-derived Schiff base and its transition metal complexes (Co, Ni, Cu, and Zn) against quite a few human cancerous cell lines, such as HepG2 liver cancer, HCT116 colorectal cancer, A549 lung carcinoma, and MCF-7 breast cancer cell lines, taking doxorubicin as the standard [11]. Of all complexes, when compared to untreated, the Zn(II) complex displayed significant activity for inhibiting the growth of HepG2, HCT116, A549, and MCF7 cell lines by a factor of 88, 70, 75, and 70, respectively. Numerous studies have been conducted in an effort to find out the improved anticancer activities of HSBs containing varied heterocycles and their several metal complexes [12–18].

A series of Cu(II) complexes were obtained from N-alkyl-substituted benzimidazole-based Schiff bases and investigated for their anticancer activities against lung (A-549), breast (MDA-MB-231), and cervical (HeLa) cancer cell lines [19]. *In vitro* cytotoxicity results revealed that the complexes act as effective drug candidates

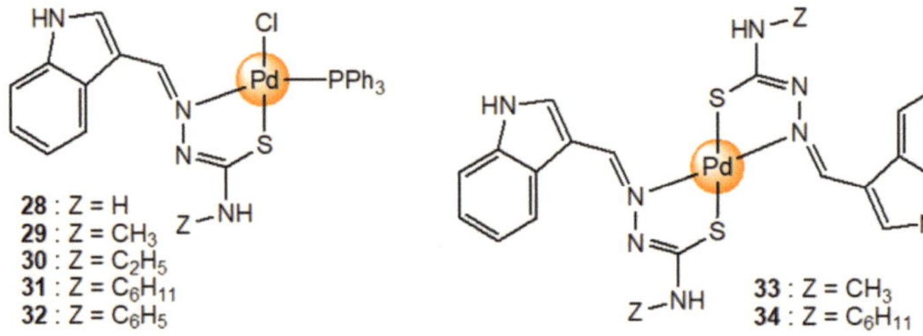

Figure 6.
Structure of indole-3-carbaldehyde thiosemicarbazone-derived palladium complexes.

against these cancer cell lines in contrast to the usually used cisplatin. A typical series of Ruthenium(II) (Ru(II)) arene HSB complexes were synthesized and assessed for their anticancer activities [20]. These complexes exhibited significant cytotoxic effects on several cancer cell lines, relying on their structure and the presence of diverse hydrophobic moieties. Furthermore, the condensation reaction between 3-(2-aminoethylamino)quinoxalin-2(1H)-one and o-vanillin resulted in the formation of quinoxalinedione-tethered Schiff base ligand, and this ligand was reacted with different metal ions, such as Cu(II), Ni(II), Co(II), and Zn(II), to form respective metal chelate complexes [21]. Upon evaluation of their anticancer properties against a human cervical cancer cell line (HeLa), it was revealed that the metal chelates perform as potential anticancer agents compared to their ligand systems.

3.2 Antimicrobial activity

The possibility of increasing mortality due to infectious diseases arises when the bacteria have multiple resistances to antibiotics. Since most of the bacteria have already adopted the situation that an antibacterial drug generates within the body, the effectiveness of the drugs against bacteria has relatively gone down. As a result, the research and development of more effective novel antibacterial drugs are undoubtedly an urgent therapeutic need [22]. In this context, Schiff base compounds have emerged as promising antibacterial agents, and their antimicrobial properties could be enhanced by metal complexations. The reason for the improved efficacy of metal complexes can be described with the help of Overtone's concept and Tweedy's chelation theory [23, 24]. According to Overtone's concept of cell permeability, the lipid membrane of the cell basically favors the passage of merely the lipid soluble constituents as liposolubility is a significant factor, which controls the activity. On chelation, the polarity of the metal ion gets reduced largely through the delocalization of electrons over chelate ring and the partial sharing of positive charge upon metal ion with donor groups, thereby enhancing the lipophilicity of the complexes. Thus, the increased lipophilicity helps metal complex to enter into the lipid membrane easily (a schematic representation is given in **Figure 7**) and restricts further multiplication of microorganisms by blocking metal-binding sites of concerned enzymes [25, 26].

Shanty et al. synthesized a series of HSBs and investigated their antibacterial activities against *Salmonella typhi*, *Bacillus coagulans*, *Bacillus pumilus*, *Escherichia coli*, *Bacillus circulans*, *Pseudomonas*, *Clostridium*, and *Klebsiella pneumoniae* [27]. All compounds exhibited low to moderate cytotoxic effect, except in the case of *Escherichia coli*. Singh et al. prepared a variety of Zn(II) complexes (35–46) (**Figure 8**) with HSBs derived from 2-hydrazino-5-[substituted phenyl]-1,3,4-thiadiazole and

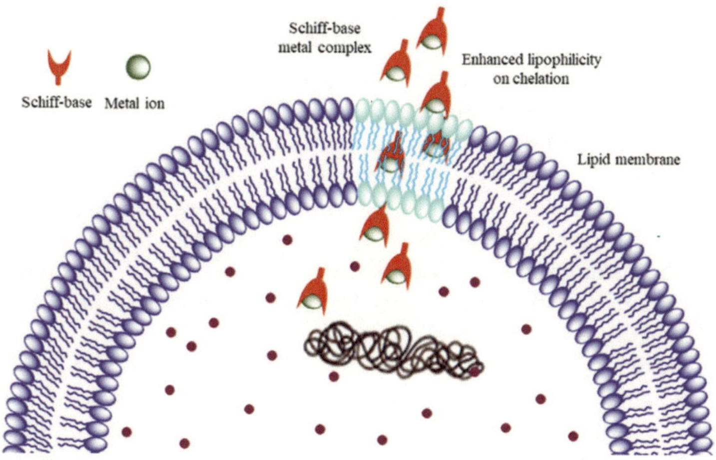

Figure 7.
Schematic representation of easy passage of metal complexes into the bacterial lipid membrane after enhanced lipophilicity of metal complexes upon chelation.

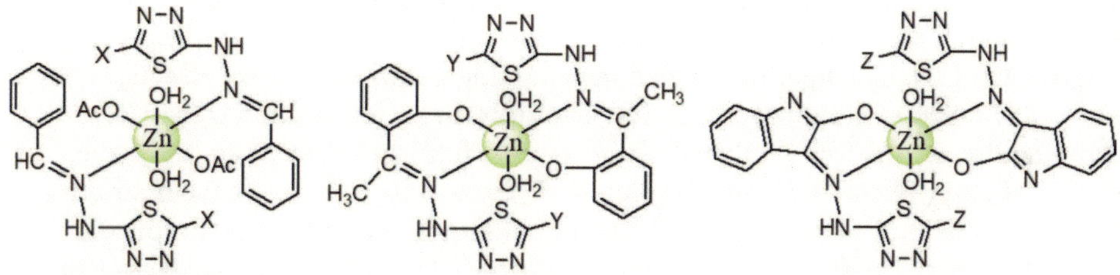

35 : X = C_6H_5
36 : X = 2-ClC_6H_4
37 : X = 4-ClC_6H_4
38 : X = 4-$NO_2C_6H_4$

39 : Y = C_6H_5
40 : Y = 2-ClC_6H_4
41 : Y = 4-ClC_6H_4
42 : Y = 4-$NO_2C_6H_4$

43 : Z = C_6H_5
44 : Z = 2-ClC_6H_4
45 : Z = 4-ClC_6H_4
46 : Z = 4-$NO_2C_6H_4$

Figure 8.
Zn(II) complexes (35–46) with HSBs derived from 2-hydrazino-5-[substituted phenyl]-1,3,4-thiadiazole and benzaldehyde/2-hydroxyacetophenone/indoline-2,3-dione.

2-hydroxyacetophenone/benzaldehyde/indoline-2,3-dione components [28]. They measured the antimicrobial properties of these compounds against bacteria (*Bacillus subtilis* and *Escherichia coli*) and fungi (*Colletotrichum falcatum*, *Aspergillus niger*, *Fusarium oxysporum*, and *Curvularia pallescens*). While antibacterial activity was measured in three respects (percentage growth inhibition, minimum inhibitory concentration, and zone of inhibition), the antifungal activity was measured only in terms of percentage growth inhibition. It revealed that, in contrast to corresponding Schiff bases, their Zn(II) complexes exhibit significantly higher antimicrobial activities. They also observed that both antibacterial and antifungal activities are enhanced with increasing chelation. The 2-chloro-substituted compounds display more activity than other substituted compounds. The compound 44 was found to be more active against all bacteria and fungi, possibly due to the presence of an additional heterocyclic ring (indoline-2,3-dione). Thus, the structure-property relationship of the ligands revealed that the additional indoline-2,3-dione ring significantly increases the pharmacological potency of Zn(II) complex 44 making it more active against both bacteria. In particular, the zone of inhibition measurement corresponding to *Bacillus subtilis* revealed that complex 44 has almost same zone (19 mm) of the standard drug streptomycin (20 mm) [28].

With an aim to explore the impact of ligand chelation with metal atoms, Karekal et al. synthesized a biologically dynamic HSB 47 from the reaction of 5-chloro-3-phenyl-1H-indole-2-carboxyhydrazide with 3-formyl-2-hydroxy-1H-quinoline, and the obtained Schiff base was complexed with Cu(II), Co(II), Ni(II), Zn(II), Cd(II), and Hg(II) ions with their proper characterizations (**Figure 9**) [29]. Antibacterial and antifungal properties of these complexes 48–53 were studied by checking their minimum inhibitory concentration (MIC). The MIC values were found to be in the ranges of 12.50–75 μg mL^{-1} and 12.50–50 μg mL^{-1} for bacterial and fungal species, respectively. The antimicrobial activity of the HSB 47 is enhanced upon complexation with metal ions mostly Cu(II), Co(II), and Ni(II), and establishing themselves essential for the growth-inhibitor effect [29].

Heterocyclic Schiff bases derived from 3-amino-2-methyl-4(3H)-quinazolinone and substituted aldehydes were combined with Pd(II) ions to get novel mononuclear Pd(II) complexes. All the ligands and their Pd(II) complexes were found to exhibit promising activity against the bacterial species, such as *B. subtilis*, *Staphylococcus aureus*, and *E. coli* [30]. Triazole-derived different HSB ligands 54–56 (**Figure 10**) were synthesized by the condensation reaction of 3-amino-1H-1,2,4-triazole with pyrrole-2-carboxaldehyde, 4-bromothiophene-2-carboxaldehyde, and 5-iodo-2-hydroxybenzaldehyde, respectively. All these ligands were reacted with Co(II), Ni(II), Cu(II), and Zn(II) ions to generate corresponding metal complexes 57–68 (**Figure 10**) [31]. The ligands showed varying degrees of inhibitory effects, with 54 displaying considerable activity (66% inhibition as determined by MIC). However, upon antimicrobial evaluation against four different Gram-negative, i.e., *E. coli, Shigella sonnei, Pseudomonas aeruginosa*, and *Salmonella typhi*, and two Gram-positive, i.e., *B. subtilis* and *S. aureus* bacterial species, metal complexes 58, 62, 64, 66, 67, and 68 showed effectual activity (54–82%) against all experimental bacterial strains [31]. In addition, the agar-well diffusion method was used to realize the antifungal activity of metal complexes against fungal species, i.e., *Candida albicans*, *Trichophyton longifusus*, *Microsporum canis*, *Aspergillus flavus*, *Fusarium*

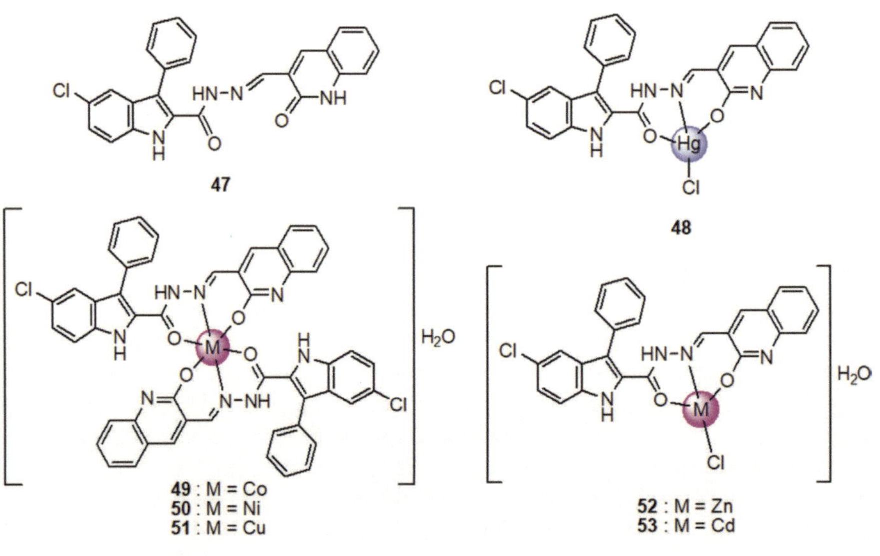

Figure 9.
Structure of hydrazone-type ligand 47 and its bioactive metal complexes 48–53.

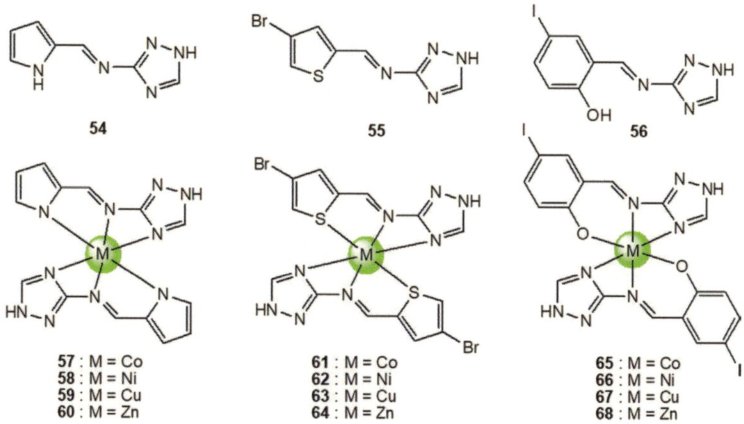

Figure 10.
Molecular structures of triazole-derived HSBs 54–56 and their metal complexes 57–68.

solani, and *Candida glabrata* [31]. The metal complexes showed significant activity, as evidenced by their higher zones of inhibition with respect to their parent ligands. In particular, the complex 58 exhibited effective activity with 73% inhibition against *C. glabrata*.

Ali et al. synthesized a number of chitosan-based Schiff bases by reacting different heterocyclic compounds, namely, 1,3-dimethyl-2,4,6-trioxohexahydropyrimidine-5-carbaldehyde, 3-acetyl-2H-chromen-2-one, 5-chloro-3-methyl-1-phenyl-1H-pyrazole-4-carbaldehyde, and 4-oxo-4H-chromene-3-carbaldehyde, with chitosan using two different approaches, namely, ultrasound and thermal [32]. These Schiff bases display thermal stability more than pure chitosan. While testing their antimicrobial activities against Gram-positive (*S. aureus* and *Enterococcus faecalis*) and Gram-negative (*E. coli* and *Salmonella typhimurium* bacteria, it showed an interesting fact that the Schiff base obtained through ultrasound method exhibited a greater degree of antimicrobial activity over the Schiff base produced through the thermal approach. The *in vitro* antimicrobial properties of the heteroleptic metal complexes containing HSB moiety were also studied by Casanova et al. [33]. Antimicrobial studies of metal (Cd and Ni) complexes display significantly higher selectivity against common Gram-positive bacteria.

3.3 Catalytic activity

Schiff bases act as potential catalysts for various organic and inorganic transformations with enhanced yields and product selectivity. The unique aspects, such as facile synthesis, structural diversity, and thermal stability, have made Schiff base ligands as well as their metal complexes more suitable for catalytic applications even in the high temperature processes, for which these ligands are extensively studied. Zhang et al. explored the ring-opening polymerization of ε-caprolactone to polycaprolactone using neodymium complex as single component catalyst that contains an aromatic HSB containing thiazole unit (**Figure 11**). The conversion of ε-caprolactone monomer was found to be as high as 94% after being polymerized for 1 hour. As investigated, the polymerization of ε-caprolactone is accompanied by a coordination anionic mechanism that involves ring opening *via* acyl-oxygen bond cleavage [34].

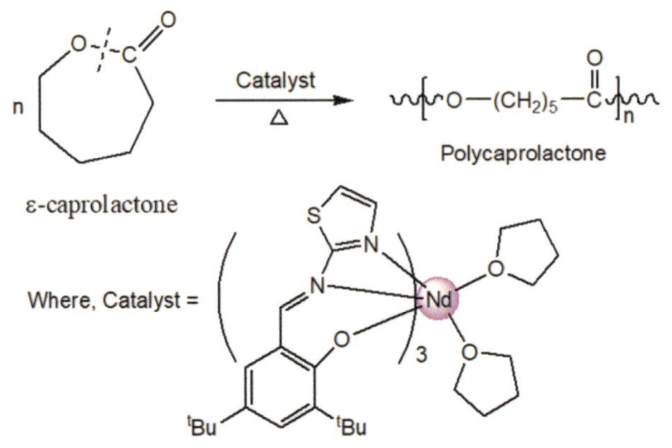

Figure 11.
Ring-opening polymerization of ε-caprolactone using HSB neodymium complex as catalyst.

One of the important reactions that find excessive applications in chemical industry is the hydrogenation of aromatic compounds. In particular, the conversion of benzene to cyclohexene is of great value in the industry, because the cyclohexene is used as a precursor for the formation of adipic acid and caprolactam, and both are found to have been used as intermediates in the synthesis of Nylon 6 and Nylon 66 [35, 36]. In contrast to simple olefins, the catalytic reduction of benzene requires more severe conditions. Arun et al. prepared two HSB metal complexes 69 and 70, and both the complexes were shown to be catalytically excellent toward the hydrogenation of benzene (**Figure 12**) [37]. Compared to 69, the catalyst 70 gives a higher conversion rate under identical experimental conditions. Interestingly, the catalyst 70 is more selective toward cyclohexane, while the catalyst 69 is more selective for cyclohexene.

A unique class of ligands, which revolutionized the field of homogeneous catalysis, is N-heterocyclic carbene (NHC). This is because, NHCs are very good σ-donor ligands and have proven to be good stabilizing ligands to a wide variety of metal ions. These ligands serve as reliable alternatives to traditional ligands, namely, amines and phosphines [38]. Exploiting the unique electron-donating properties of NHC, Abubakar et al. synthesized a trinuclear Ni(II) complex that contains a NHC functionalized Schiff base. This particular complex was found to have excellent

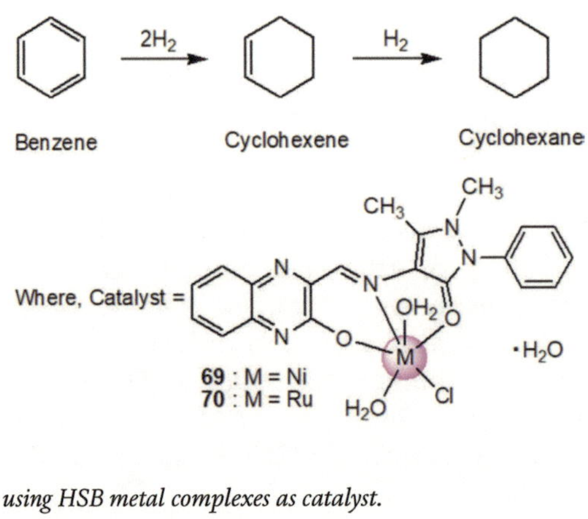

Figure 12.
Hydrogenation of benzene using HSB metal complexes as catalyst.

catalytic property in the *transfer hydrogenation* reaction of a range of aromatic and aliphatic ketones (**Figure 13**), at a very low catalyst concentration of 0.1 mol% [38]. As revealed, there were quantitative conversions of aromatic ketones to corresponding alcohols in the presence of isopropyl alcohol, and these values were found to be similar to the values obtained by means of a more expensive Ru-complex catalyst containing no heterocyclic moieties at a concentration of 0.25 mol% [39].

It has been observed that half-sandwich Ru-complexes play an important role as catalysts in the synthesis of valuable organic compounds. In particular, half-sandwich Ru-complexes containing HSB ligands with oxygen-nitrogen donor centers effectively catalyze the transfer hydrogenation of a wide range of ketones. Several half-sandwich Ru-complexes (71–76) bearing HSB ligands, which in turn are being used as catalysts for transfer hydrogenation of ketones, are shown in **Figure 14** [40, 41].

The compound *p*-aminophenol is quite useful and has industrial importance due to its extensive applications in different fields, such as making antipyretic and analgesic drugs, corrosion inhibitors, hair-dyeing agents, photographic developers, and anticorrosion lubricants [42]. Recently, two Ni(II) complexes with a pyrazole-based HSB ligand have been reported by Mandal et al. [43], and both the complexes were found to be remarkably capable of converting *p*-nitrophenol to *p*-aminophenol in the presence of sodium borohydride as reducing agent. Santa Barbara Amorphous No. 15 (SBA-15) is an interesting mesoporous silica material, and it exhibits promising catalytic activity for wider applications having unique properties of highly ordered mesopores, thick wall, hydrothermal stability, large surface area, and enormous pore volumes [44]. Since it lacks functionality, heteroatoms and organic functional groups can be introduced, either by direct or by post-synthesis methods to modify

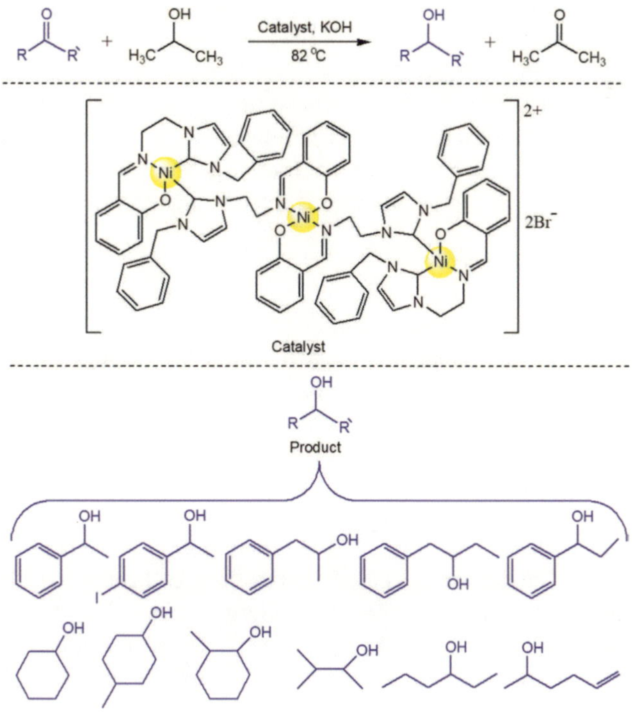

Figure 13.
Catalytic transfer hydrogenation of ketones to corresponding alcohols.

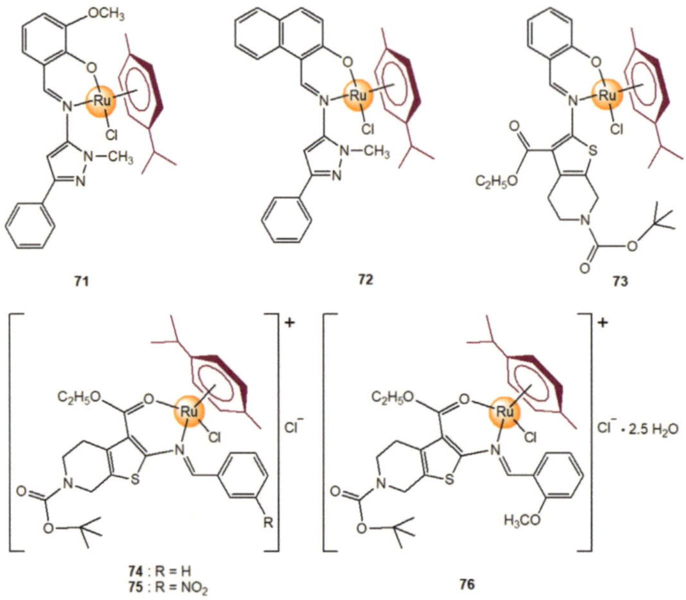

Figure 14.
Molecular structures of half-sandwich Ru-complexes 71–76 bearing HSB ligands.

its functionality, thereby enhancing the catalytic activity [45]. In view of this, Yari et al. prepared SBA-15-supported cobalt complex as a novel heterogeneous nanocatalyst comprising N–O chelating HSB ligand [46]. Initially, they synthesized HSB by reacting pyridoxal 5′-phosphate or PLP (biologically active form of vitamin B_6) with 3-(aminopropyl)-triethoxysilane, and complexed with Co(II) ion, which on subsequent grafting to SBA-15 afforded Co(II)-PLP-HSB/SBA-15 nanocatalyst. This particular nanocatalyst is suitable for the synthesis of several benzothiazole heterocycles (77–92) by the cyclo-condensation of aryl-aldehydes with 2-aminothiophenol having yields good to excellent under green conditions [46]. The catalyst is active, stable, and can be reused without leaching of metal ions into the reaction medium (**Figure 15**).

3.4 Miscellaneous applications

In addition to the applications mentioned at earlier sections, HSBs and their related metal complexes also find their uses in other fields, and a few instances are described here in brief. By and large, the most extensively used forms of energy derived from nonrenewable sources include coal, natural gas, and petroleum, and in fact, these sources of energy become key factors of environmental pollution creating an unhealthy atmosphere for natural habitats. In such a scenario, it is crucial to find alternate eco-friendly sources of energy which will be capable of producing adequate power to meet the world's energy crisis. Thus, solar energy is considered one of the sensible choices, and for this, solar cell or photovoltaic cell acts as a unique device that converts sun's light energy into electrical energy. Researchers have been continuously putting their utmost effort to improve the efficiency of a solar cell, and the use of HSB compounds has played a great role in this regard. For example, Chen et al. synthesized polymeric HSB with thiophene units, and the polymer was employed as an active layer for polymer solar cell devices. The conductivity of polymers was substantially improved under natural light condition, indicating the possession of

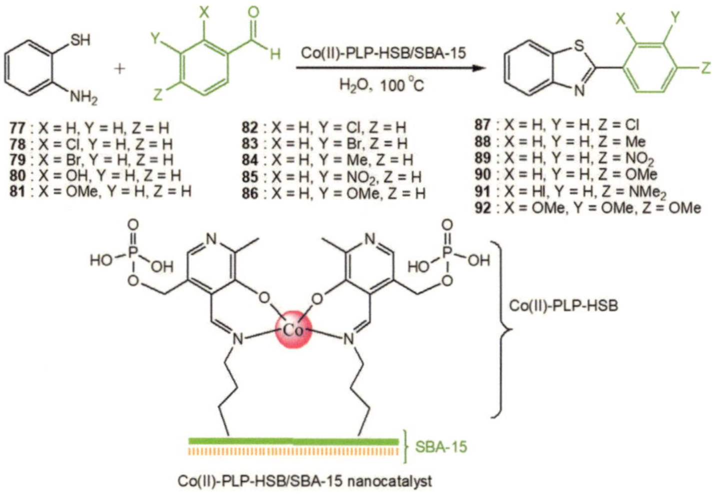

Figure 15.
Preparation of benzothiazole heterocycles using Co(II)-PLP-HSB/SBA-15 nanocatalyst.

photoelectric conversion efficiency with a potential in conjugated polymer photovoltaic materials [47]. Quinoline-derived Schiff base and its chromium complex were synthesized by Salman et al., and both were used to make nano-films on Si layer *via* drop casting technique [48]. The accrued efficiency of the fabricated inorganic silicon solar cell was found to be higher than that of the organic solar cell. Similarly, the pyrazole-derived HSB ligand and its nickel complex were utilized to make the thin films for improving the solar cell efficiency [49]. Interestingly, the asymmetric fluorescence benzimidazole complexes were obtained from the half-condensed asymmetrical HSB ligand, [N^1-((quinoline-2-yl) methylene) N^2-(H-(quinoline-2-yl) methylenehydroxy)o-phenylenediamine], by means of a metal(II)-induced method [50]. These complexes were used as cosensitizers in dye-sensitized solar cell to enhance its performance.

Benzanthrone is a polycyclic aromatic hydrocarbon and its derivatives find multiple useful applications in the chemical industry. Orlova et al. synthesized novel benzanthrone azomethines by condensing 3-aminobenzanthrone with a series of heterocyclic aldehydes [51]. The obtained benzanthrone derivatives were reduced to corresponding thermally stable amines with higher luminescence efficiency. These compounds absorb light at 420–525 nm with quite large Stokes shifts and emit at 500–660 nm. Pertinent results suggested that the emission is sensitive to the polarity of solvent exhibiting negative fluorosolvatochromism for azomethines and positive fluorosolvatochromism for reduced azomethines [51]. The high fluorescence in the red region intensifies the possibilities of practical application of the prepared amines that are promising as luminescent dyes.

Corrosion inhibitors are very indispensable for protecting metals from destructive chemical reactions that occurred in the presence of moisture, humidity, and acidic conditions [52, 53]. Generally, inhibitors possessing increased basicity centers, such as N, O, and S, unsaturated bonds, and planar-conjugated aromatic compounds are endorsed as active corrosion inhibitors [54, 55]. Because of the presence of the imine bond, the use of Schiff bases as corrosion inhibitors in metals and alloys has been the major motivation [56, 57]. Boulechfar et al. synthesized 2-furaldehyde semicarbazone HSB metal complexes of Mn(II), Co(II), and Zn(II), and the corrosion inhibitions of these complexes against the XC38 carbon steel were evaluated in 1 M hydrochloric acid (HCl) [58]. Among these complexes, Mn-complex was found to have the highest inhibitory effect at around 91%. These findings are highly important in demonstrating the effectiveness of the HSB-based metallic complex inhibitors in protecting the surfaces of metals, storage tanks, pipelines, and other industrial equipment that are exposed to corrosive environs. The incorporation of these inhibitors into various materials such as paint and plastics will enhance their resistance to corrosion significantly. Recently, the dipeptide HSBs synthesized by S. Satpati et al. [59] have been found to possess excellent anticorrosion properties. These compounds exhibit outstanding efficiency in preventing corrosion of mild steel.

The development of chemosensors for selective sensing of analytes continues to be one of the most essential research areas of supramolecular chemistry. A chemosensor is a molecule whose interaction with the analyte produces a detectable change or signal. A series of studies have revealed that colorimetric and fluorescent chemosensors are of crucial molecules for sensing ionic or neutral species through changes in color or fluorescence. Krishna et al. in 2019 synthesized a number of isatin-based HSBs as colorimetric chemosensors [60]. Out of many analytes such as AsO_2^-, Hg^{2+}, Mn^{2+}, Fe^{2+}, Co^{2+}, Ni^{2+}, Zn^{2+}, Al^{3+}, Pb^{2+}, Ag^+, Cd^{2+}, Cr^{3+}, Ca^{2+}, Na^+, Mg^{2+}, K^+, and Cu^{2+} that were subjected to detection, the chemosensors selectively detected the arsenite and Hg^{2+} ions only through color change visible to the naked eye [60]. Because of the facile and selective detection of both toxic ions, the strategy could be extended to both industrial applications and the environmental monitoring against hazardous metal ions. Moreover, the HSB comprising indole-triazole motif provides a platform for effective binding unit to recognize a variety of cations and anions, and serves as well-known fluorescence chemosensors [16]. In 2018, Singh et al. synthesized indole-triazole HSB as chemosensor for the selective sensing of Ni^{2+} and Cu^{2+} ions over a wide range of metal ions [61]. Novel triazole-coumarin HSB, a fluorescent chemosensor, was prepared by Wang et al. [62] and this particular chemosensor was found to be suitable for selective sensing of Cu^{2+} ions. Furthermore, the bis(phenyl)fluorene and carbazole appended dipodal HSBs were synthesized and evaluated as efficient fluorescent chemosensors for detection of Sn^{2+} ions [63]. The study revealed that both nitrogen atoms in the two imine groups of HSB might lead to electron transfer to vacant orbitals of Sn^{2+}, resulting in C=N isomerization quenching. The fluorescence quantum yield of dipodal HSB-Sn^{2+} chelation was found to be 17.4% with an outstanding photostability under excitation at 320 nm.

Hence, the structural diversities and inherent chemical properties of HSB-based molecules provide scopes for numerous utilities in different segments. Without limiting their applications in the above-mentioned zones, these molecules too exhibit some foremost activities in the field of antimalarial, antiviral, anti-inflammatory, antioxidant, and anti-obesity that have been documented in the literature [16, 64–66]. In the contemporary period, the pharmacological applications of HSB-related compounds have excellent potential and are widely used within the pharmaceutical sector.

4. Conclusions

The unique features of HSBs and their coordination complexes with several transition metal ions offer a significant prospect for the design of novel drug molecules against a variety of cancer cell lines as well as active pathogens. Certainly, the HSB ligands have good biological properties, and the ability of heteroatoms in HSBs to form chelate complexes with pharmacologically important metal ions corresponding to cobalt, nickel, copper, and zinc is an extra advantage to their pharmacophore properties. The enhanced activity of complexes can rationally be credited to a cluster of factors, such as the occurrence of chelation, reduced polarity of metal ions, increased lipophilicity, accessibility of suitable heteroatom(s) in HSB ligand, and geometry of the complexes. The HSBs that contain N, O, and S heteroatoms owned better cytotoxicity against microbial strains and their activities have still been nurtured radically after metalation. Moreover, the chemical and thermal stabilities of HSB complexes make them efficient catalysts for a wide range of important organic transformation reactions that include ring-opening polymerization, hydrogenation of benzene, transfer hydrogenation of ketones, cyclo-condensation of aryl-aldehydes with 2-aminothiophenol, etc. The study can further be stimulated by exploiting the chemistry of uranyl complexes corresponding to their extensive HSB ligands for the catalytic and therapeutic studies due to their intriguing structural diversities along with an unprecedented complex core. The inexorable research activity focused on Schiff base compounds and their prevalent applications can be expected to continue, even in the future to yield more attractive results in terms of both novelty and function. The insights provided in this chapter can help researchers to rationally design novel HSB derivatives as well as pertinent metal complexes for suitable applications in multiple domains.

Author details

Nabakrushna Behera*, Tankadhar Behera, Jyotiprabha Rout and Sasmita Moharana
School of Chemistry, Sambalpur University, Sambalpur, Odisha, India

*Address all correspondence to: nkbehera@suniv.ac.in

References

[1] Tidwell TT. Hugo (Ugo) Schiff, Schiff bases, and a century of β-lactam synthesis. Angewandte Chemie, International Edition. 2008;**47**:1016-1020

[2] Akhtar W, Khan MF, Verma G, Shaquiquzzaman M, Rizvi MA, Mehdi SH, et al. Therapeutic evolution of benzimidazole derivatives in the last quinquennial period. European Journal of Medicinal Chemistry. 2017;**126**:705-753

[3] Mahire VN, Mahulikar PP. Facile one-pot clean synthesis of benzimidazole motifs: Exploration on bismuth nitrate accelerated subtle catalysis. Chinese Chemical Letters. 2015;**26**:983-987

[4] Hranjec M, Starčević K, Pavelić SK, Lučin P, Pavelić K, Zamola GK. Synthesis, spectroscopic characterization and antiproliferative evaluation in vitro of novel Schiff bases related to benzimidazoles. European Journal of Medicinal Chemistry. 2011;**46**:2274-2279

[5] Ebenezer O, Oyetunde-Joshua F, Omotoso OD, Shapi M. Benzimidazole and its derivatives: Recent Advances (2020-2022). Results in Chemistry. 2023;**5**:100925

[6] Priya B, Utreja D, Sharma S, Kaur G, Madhvi. Synthesis and antimicrobial activities of indole-based Schiff bases and their metal complexes: A review. Current Organic Chemistry. 2023;**27**:941-961

[7] Mohanambal D, Sridevi G, Antony SA, Angayarkani R. A recent review in applications of heterocyclic compounds as antimicrobial agent. Asian Journal of Pharmaceutical and Clinical Research. 2018;**11**:66-74

[8] Ghaidan AF, Faraj FL, Abdulghany ZS. Synthesis, characterization and cytotoxic activity of new indole Schiff bases, derived from 2-(5-chloro-3,3-dimethyl-1,3-dihydro-indol-2-ylidene)-malonaldehyde with substituted aniline. Oriental Journal of Chemistry. 2018;**34**:169-181

[9] Kunj MT, Patel UP, Dabhi RC, Maru JJ. Synthesis, computational insights, and anticancer activity of novel indole-Schiff base derivatives. Russian Journal of Bioorganic Chemistry. 2022;**48**:601-608

[10] Haribabu J, Tamizh MM, Balachandran C, Arun Y, Bhuvanesh NSP, Endo A, et al. Synthesis, structures and mechanistic pathways of anticancer activity of palladium(II) complexes with indole-3-carbaldehyde thiosemicarbazones. New Journal of Chemistry. 2018;**42**:10818-10832

[11] Al-Shemary RK, Mohapatra RK, Kumar M, Sarangi AK, Azame M, Tulif HS, et al. Synthesis, structural investigations, XRD, DFT, anticancer and molecular docking study of a series of thiazole based Schiff base metal complexes. Journal of Molecular Structure. 2023;**1275**:134676

[12] Tyagi P, Tyagi M, Agrawal S, Chandra S, Ojha H, Pathak M. Synthesis, characterization of 1,2,4-triazole Schiff base derived 3d-metal complexes: Induces cytotoxicity in HepG2, MCF-7 cell line, BSA binding fluorescence and DFT study. Spectrochimica Acta, Part A: Molecular and Biomolecular Spectroscopy. 2017;**171**:246-257

[13] Lukmantara AY, Kalinowski DS, Kumar N, Richardson DR. Structure–activity studies of 4-phenyl-substituted 2′-benzoylpyridine thiosemicarbazones

with potent and selective anti-tumour activity. Organic & Biomolecular Chemistry. 2013;**11**:6414-6425

[14] Pettinari R, Pettinari C, Marchetti F, Clavel CM, Scopelliti R, Dyson PJ. Cytotoxicity of ruthenium–arene complexes containing β-ketoamine ligands. Organometallics. 2013;**32**:309-316

[15] Shakir M, Azam M, Ullah MF, Hadi SM. Synthesis, spectroscopic and electrochemical studies of N,N-bis[(E)-2-thienylmethylidene]-1,8-naphthalenediamine and its Cu(II) complex: DNA cleavage and generation of superoxide anion. Journal of Photochemistry and Photobiology. B. 2011;**104**:449-456

[16] Sanjurani T, Barman P. The versatile nature of indole containing Schiff bases: An overview. Polyhedron. 2024;**249**:116779

[17] Malik MA, Dar OA, Gull P, Wani MY, Hashmi AA. Heterocyclic Schiff base transition metal complexes in antimicrobial and anticancer chemotherapy. Medicinal Chemistry Communications. 2018;**9**:409-436

[18] Beč A, Cindrić M, Persoons L, Banjanac M, Radovanović V, Daelemans D, et al. Novel biologically active N-substituted benzimidazole derived Schiff bases: Design, synthesis, and biological evaluation. Molecules. 2023;**28**:3720

[19] Paul A, Anbu S, Sharma G, Kuznetsov ML, Koch B, Guedes da Silva MFC, et al. Synthesis, DNA binding, cellular DNA lesion and cytotoxicity of a series of new benzimidazole-based Schiff base copper(II) complexes. Dalton Transactions. 2015;**44**:19983-19996

[20] Chow MJ, Babak MV, Wong DYQ, Pastorin G, Gaiddon C, Ang WH. Structural determinants of p53-independence in anticancer ruthenium-arene Schiff-base complexes. Molecular Pharmaceutics. 2016;**13**:2543-2554

[21] Chellaian JD, Johnson J. Spectral characterization, electrochemical and anticancer studies on some metal(II) complexes containing tridentate quinoxaline Schiff base. Spectrochimica Acta, Part A: Molecular and Biomolecular Spectroscopy. 2014;**127**:396-404

[22] Pervaiz M, Quratulain R, Ejaz A, Shahin M, Saeed Z, Nasir S, et al. Thiosemicarbazides, 1,3,4 thiadiazole Schiff base derivatives of transition metal complexes as antimicrobial agents. Inorganic Chemistry Communications. 2024;**160**:111856

[23] Tanja S, Enisa S, Braho L. Second international electronic conference on medicinal chemistry. Pharmaceuticals (Basel). 2016;**10**(1):1-30

[24] Tweedy BG. Plant extracts with metal ions as potential antimicrobial agents. Phytopathology. 1964;**55**:910-918

[25] Beyene BB, Mihirteu AM, Ayana MT, Yibeltal AW. Synthesis, characterization and antibacterial activity of metalloporphyrins: Role of central metal ion. Results in Chemistry. 2020;**2**:100073

[26] Mishra AP, Mishra R, Jain R, Gupta S. Synthesis of new VO(II), Co(II), Ni(II) and Cu(II) complexes with isatin-3-chloro-4-floroaniline and 2-pyridinecarboxylidene-4-aminoantipyrine and their antimicrobial studies. Mycobiology. 2012;**40**:20-26

[27] Shanty AA, Philip JE, Sneha EJ, Kurup MRP, Balachandran S, Mohanan PV. Synthesis, characterization

and biological studies of Schiff bases derived from heterocyclic moiety. Bioorganic Chemistry. 2017;**70**:67-73

[28] Singh AK, Pandey OP, Sengupta SK. Synthesis, spectral and antimicrobial activity of Zn(II) complexes with Schiff bases derived from 2-hydrazino-5-[substituted phenyl]-1,3,4-thiadiazole and benzaldehyde/2-hydroxyacetophenone/indoline-2,3-dione. Spectrochimica Acta, Part A: Molecular and Biomolecular Spectroscopy. 2013;**113**:393-399

[29] Karekal MR, Biradar V, Bennikallu M, Mathada H. Synthesis, characterization, antimicrobial, DNA cleavage, and antioxidant studies of some metal complexes derived from Schiff base containing indole and quinoline moieties. Bioinorganic Chemistry and Applications. 2013;**2013**:315972

[30] Shiva K, Shiva L, Chandan S, Kumar RMN, Revanasiddappa HD. Palladium(II) complexes as biologically potent metallo-drugs: Synthesis, spectral characterization, DNA interaction studies and antibacterial activity. Spectrochimica Acta, Part A: Molecular and Biomolecular Spectroscopy. 2013;**107**:108-116

[31] Hanif M, Chohan ZH. Design, spectral characterization and biological studies of transition metal(II) complexes with triazole Schiff bases. Spectrochimica Acta, Part A: Molecular and Biomolecular Spectroscopy. 2013;**104**:468-476

[32] Ali M, Sholkamy EN, Alobaidi AS, Al-Muhanna MK, Barakat A. Synthesis of Schiff bases based on chitosan and heterocyclic moiety: Evaluation of antimicrobial activity. ACS Omega. 2023;**8**:47304-47312

[33] Casanova I, Durán ML, Viqueira J, Sousa-Pedrares A, Zani F, Realc JA, et al. Metal complexes of a novel heterocyclic benzimidazole ligand formed by rearrangement-cyclization of the corresponding Schiff base. Electrosynthesis, structural characterization and antimicrobial activity. Dalton Transactions. 2018;**47**:4325-4340

[34] Zhang L, Ni X-F, Sun W-L, Shen Z-Q. Heterocyclic Schiff base neodymium complex as catalyst for ring-opening polymolarization of ε-caprolactone. Chinese Journal of Polymer Science. 2008;**26**:475-479

[35] Weissermel K, Arpe H-J. Industrial Organic Chemistry. New York: VCH; 1993

[36] Parshall GW, Ittel SD. The applications and chemistry of catalysis by soluble transition metal complexes. In: Homogeneous Catalysis. 2nd ed. New York: Wiley; 1992

[37] Arun V, Sridevi N, Robinsona PP, Manjua S, Yusuff KKM. Ni(II) and Ru(II) Schiff base complexes as catalysts for the reduction of benzene. Journal of Molecular Catalysis A: Chemical. 2009;**304**:191-198

[38] Abubakar S, Ibrahim H, Bala MD. Transfer hydrogenation of ketones catalyzed by a trinuclear Ni(II) complex of a Schiff base functionalized N-heterocyclic carbene ligand. Inorganica Chimica Acta. 2019;**484**:276-282

[39] Biswas S, Sarkar D, Roy P, Mondal TK. Synthesis and characterization of a ruthenium complex with bis (diphenylphosphino)propane and thioether containing ONS donor ligand: Application in transfer hydrogenation of ketones. Polyhedron. 2017;**131**:1-7

[40] Buldurun K, Özdemir M. Ruthenium(II) complexes with pyridine-based Schiff base ligands: Synthesis,

structural characterization and catalytic hydrogenation of ketones. Journal of Molecular Structure. 2020;**1202**:127266

[41] Ramesh M, Venkatachalam G. Half–sandwich (η6–p–Cymene) ruthenium(II) complexes bearing 5–amino–1–methyl–3–phenylpyrazole Schiff base ligands: Synthesis, structure and catalytic transfer hydrogenation of ketones. Journal of Organometallic Chemistry. 2019;**880**:47-55

[42] da Silva AGM, Rodrigues TS, Macedo A, da Silva RTP, Camargo PHC. An undergraduate level experiment on the synthesis of Au nanoparticles and their size dependent and their size-dependent optical and catalytic properties. Quimica Nova. 2014;**37**:1716-1720

[43] Mandal S, Sarkar M, Denrah S, Bagchi A, Biswas A, Cordes DB, et al. Catalytic and anticancer activity of two new Ni(II) complexes with a pyrazole based heterocyclic Schiff-base ligand: Synthesis, spectroscopy and X-ray crystallography. Journal of Molecular Structure. 2023;**1287**:135648

[44] Rahmat N, Abdullah AZ, Mohamed AR. A review: Mesoporous santa barbara amorphous-15, types, synthesis and its applications towards biorefinery production. American Journal of Applied Sciences. 2010;**7**:1579-1586

[45] Huirache-Acuña R, Nava R, Peza-Ledesma CL, Lara-Romero J, Alonso-Núñez G, Pawelec B, et al. SBA-15 mesoporous silica as catalytic support for hydrodesulfurization catalysts–Review. Materials. 2013;**6**:4139-4167

[46] Yari H, Dehkharghani RA, Bardajee GR, Akbarzadeh-T N. Synthesis, characterization, and applications of novel Co(II)-pyridoxal phosphate-Schiff base/SBA-15 as a nanocatalyst for the green synthesis of benzothiazole heterocycles. Journal of the Chinese Chemical Society. 2020;**67**:1490-1500

[47] Chen W, Chui P, Ma A, Luo C, Gu Y, Mi Q. Synthesis of polythiophene derivatives with Schiff base groups and use as an active layer in polymer solar cell devices. Materials Science Forum. 2015;**815**:668-674

[48] Salman AT, Ismail AH, Rheima AM, Abd AN, Habubi NF, Abbas ZS. Nano-synthesis, characterization and spectroscopic studies of chromium (III) complex derived from new quinoline-2-one for solar cell fabrication. Journal of Physics: Conference Series. 2021;**1853**:012021

[49] Ismail AH, Al-Zaidi BH, Abd AN, Habubi NF. Synthesis and characterization of novel thin films derived from pyrazole 3 one and its metal complex with bivalent nickel ion to improve solar cell efficiency. Chemical Papers. 2020;**74**:2069-2078

[50] Wang X, Fan R, Dong Y, Su T, Huang J, Du X, et al. Metal(II)-induced synthesis of asymmetric fluorescence benzimidazoles complexes and their dye-sensitized solar cell performance as cosensitizers. Crystal Growth & Design. 2017;**17**:5406-5421

[51] Orlova N, Nikolajeva I, Pučkins A, Belyakov S, Kirilova E. Heterocyclic Schiff bases of 3-aminobenzanthrone and their reduced analogues: Synthesis, properties and spectroscopy. Molecules. 2021;**26**:2570

[52] Kar PK, Panda AK, Behera N, Das PK, Panda AK. Evaluation of Schiff's bases synthesised from 2-hydrazinobenzimidazole as corrosion inhibitor of copper in H_2SO_4. Materials

Science: An Indian Journal (MSAIJ). 2011;7:350-352

[53] Trabanelli G. Inhibitors–An old remedy for a new challenge. Corrosion. 1991;**47**:410-419

[54] Deng S, Li X, Fu H. Two pyrazine derivatives as inhibitors of the cold rolled steel corrosion in hydrochloric acid solution. Corrosion Science. 2011;**53**:822-828

[55] Hasanov R, Sadıkoğlu M, Bilgiç S. Electrochemical and quantum chemical studies of some Schiff bases on the corrosion of steel in H_2SO_4 solution. Applied Surface Science. 2007;**253**:3913-3921

[56] Boulechfar C, Ferkous H, Delimi A, Djedouani A, Kahlouche A, Boublia A, et al. Schiff bases and their metal complexes: A review on the history, synthesis, and applications. Inorganic Chemistry Communications. 2023;**150**:110451

[57] Nassar AM, Hassan AM, Shoeib MA, El Kmash AN. Synthesis, characterization and anticorrosion studies of new homobimetallic Co(II), Ni(II), Cu(II), and Zn(II) Schiff base complexes. Journal of Bio- and Tribo-Corrosion. 2015;**1**:19

[58] Boulechfar C, Ferkous H, Delimi A, Berredjem M, Kahlouche A, Madaci A, et al. Corrosion inhibition of Schiff base and their metal complexes with [Mn(II), Co(II) and Zn(II)]: Experimental and quantum chemical studies. Journal of Molecular Liquids. 2023;**378**:121637

[59] Satpati S, Suhasaria A, Ghosal S, Dey S, Sukul D. Interaction of newly synthesized dipeptide Schiff bases with mild steel surface in aqueous HCl: Experimental and theoretical study on thermodynamics, adsorption and anti-corrosion characteristics. Materials Chemistry and Physics. 2023;**296**:127200

[60] Krishna A, Tekuri V, Mohan M, Trivedi DR. Selective colorimetric chemosensor for the detection of Hg^{2+} and arsenite ions using isatin based Schiff's bases; DFT studies and applications in test strips. Sensors & Actuators, B: Chemical. 2019;**284**:271-280

[61] Singh G, Kalra P, Arora A, Sanchita G, Sharma A, Singh VV. Inorg, Design and synthesis of indole triazole pendant siloxy framework as a chemo sensor for sensing of Cu^{2+} and Ni^{2+}: A comparison between traditional and microwave method. Chimica Acta. 2018;**473**:186-193

[62] Wang W, Yuan W-J, Liu Q-L, Lei Y-N, Qi S, Gao Y. Selective fluorescence sensor for Cu^{2+} with a novel triazole Schiff-base derivative with coumarin units. Heterocyclic Communications. 2014;**20**:289-292

[63] Kolcu F, Çulhaoğlu S, Kaya I. Synthesis and investigation of bis(phenyl) fluorene and carbazole appended dipodal Schiff base for fluorescence sensing towards Sn(II) ion and its regioselective polymerizations. Journal of Photochemistry and Photobiology A: Chemistry. 2023;**441**:114665

[64] Fonkui TY, Ikhile MI, Ndinteh DT, Njobeh PB. Microbial activity of some heterocyclic Schiff bases and metal complexes: A review. Tropical Journal of Pharmaceutical Research. 2018;**17**: 2507-2518

[65] Alomari AA, Ibrahim MM, Mohamed ME. Schiff bases of hydroxyacetophenones and their copper(II) and nickel(II) complexes: Synthesis, antioxidant activity and theoretical study. Asian Journal of Chemistry. 2016;**28**:2505-2511

[66] Mushtaq I, Ahmad M, Saleem M, Ahmed A. Pharmaceutical significance of Schiff bases: An overview. Future Journal of Pharmaceutical Sciences. 2024;**10**:16

Chapter 5

Metal Complex

Shilpa Laxman Sangle, Rekha Barku More and Savita V. Thakare

Abstract

The scientific community has given Schiff-based metallic complexes much attention because of their distinctive properties and numerous uses in a variety of fields, including biology and industry. The nature of the ligands and metal ions influences coordination chemistry. The biological activities of these complexes include cytostatic, antiviral, antibacterial, anticancer, and antifungal effects. They also have remarkable catalytic activity. With its diverse uses, coordination complexes serve a vital role in research.

Keywords: Schiff base, metal complexes, applications, ligand, synthesis

1. Introduction

A metal complex is composed of one or more ligands, which are molecules or ions that have one or more pairs of electrons that they can share with the metal. Metal complexes can be neutral, positively charged, or negatively charged. A compound that forms a coordination complex by bonding with a central metal atom or ion is called a ligand. The ligand is an electron-rich compound that can be an anion, cation, or neutral. Ionic or covalent bonding is possible in metal-ligand complexes. Depending upon the number of donating atoms, the ligands may be bidentate, tridentate, or polydentate. The condensation of primary amines with aldehydes or ketones gives compounds that contain a carbon-nitrogen double bond with the nitrogen attached to an alkyl group. These are called Schiff bases since their synthesis was reported by Hugo Schiff [1].

Coordination chemistry is one of the most active research areas in inorganic chemistry, which involves the interaction of organic/inorganic ligands with metal centers. Numerous fields, such as dyes, colors, nuclear fuels, catalysis, toxicology, bioinorganic chemistry, medicine, ceramics, materials science, and toxicology, uses various forms of coordination compounds. The application of these complexes as biocides, catalysts, NMR shift reagents, and DNA binders is due to the inclusion of a numerous kinds of ligands [2]. Because of this work, inorganic chemists have been able to significantly advance their understanding of how to modify the concept of chemical bonding [3].

The synthesis, characterization, and uses of various ligands, including Schiff base ligands and chelates, are well documented in coordination chemistry [4]. Many coordination chemists are interested in transition metal complexes with higher nuclearity, and polytopic ligands play a key role in modifying the molecular structures of these complexes. While one-pot syntheses of polynuclear complexes are highly complex,

IntechOpen

whereas polytopic ligands are superior for the successful creation of homopolynonucleic complexes [5–10].

The development of synthetic pathways is of greatest significance in order to enable the regulated and predictable tailor-making of heterometallic systems. Among the several methodologies, the use of metal complexes as ligands has shown to be quite effective. If the complexes formed by coordination with metal ions have the tendency to coordinate further or react with other complexes, then they may act as 'metal organic ligands' [11]. In a variety of applications, polynuclear complexes are particularly helpful as magnetic materials. Magnetic materials are currently being utilized in biomedical applications, including magnetic separation of cells, DNA, proteins, and other materials, and targeted drug delivery in magnetic resonance imaging (MRI) [12–17].

1.1 Introduction to metal complexes

Over the past few decades, metal complexes synthesized from different Schiff base ligands served a significant role in coordination chemistry. Because of their simplicity in synthesis and many uses, these ligand derivatives are frequently used. Schiff base ligand metal complexes have become popular due to their remarkable biological, catalytic, and fluorescent properties. Many Schiff base metal complexes have been developed and tested for their antibacterial and antifungal qualities because Schiff bases containing azomethine or imine groups have a variety of biological activities that can advance the field of bioinorganic sciences [18–20].

Researchers are working harder to create novel antibiotic compounds because of the growing number of bacterial and fungal strains that are becoming resistant to traditional antibiotics. Schiff base has an azomethine (–C=N–) functional group, which forms bonds with two or more aromatic or heterocyclic compounds that are physiologically active to generate a variety of molecular hybrids with excellent antibacterial capabilities. These Schiff bases are chelating compounds that often create stable chelates by coordinating with metals, particularly lanthanides and d-block metals with a wide range of therapeutic applications [21, 22]. These Schiff bases form a new family of compounds with promising antibacterial and anticancer properties. Schiff base metal complexes, which are derived from halogenated salicylaldehyde, have a broad variety of applications in research involving biocatalytic, anti-HIV, antibacterial, and luminous probes for DNA and RNA breakage reactions [23–25]. These derivatized complexes are also employed as fungicides, insecticides, and flavoring agents for bouquets and liquor [26–32].

The structure and vibrational spectra of Schiff base complexes have been investigated by various methods [33–40], but few studies have investigated symmetric azine compounds and their complexes [41, 42]. Azines are important due to their analogy with Schiff bases, which are considered models of some biological systems [43–50]. Recently, the stereochemistry and stereo electronics of some substituted (E, E)-azines [51] and the crystal structure of asymmetric acetophenone azines have been studied [52] since these compounds are considered nonlinear optical materials.

2. Synthesis

Schiff base synthesis by green synthetic methods has gained significant attention in recent years because of increasing need for more environmentally friendly and sustainable chemical production methods. Traditional methods of synthesis of Schiff bases involve extreme circumstances, such as high temperatures and acidic or basic catalysts, which can produce hazardous byproducts. Green synthetic techniques, on the other hand, use softer conditions and renewable resources to prepare Schiff bases. General Schiff base formation includes ketone or aldehyde, where the >C=O group is substituted with the >C=N–R group (**Figure 1**) where R may be an alkyl, aryl, halogen, or hydrogen group. Schiff bases containing aryl substituent are significantly more stable than those with alkyl substituents [53].

Aromatic aldehydes Schiff bases are more durable because of strong conjugation, while aliphatic aldehyde Schiff bases are highly unstable and easily polymerizable. The stability of Schiff bases is also affected by the substituents, as a result, several methods for producing imines have been reported [54]. A Schiff base can be formed from an aldehyde or ketone in an acidic, basic, or heated environment. In **Figure 2**, the reversible reaction is depicted.

The formation is usually pushed to the finishing touch of these preparations of the product or elimination of water molecules, or both. Many Schiff bases can be hydrolyzed in an aqueous acidic or basic medium to return to their aldehydes or ketones and amines. The reaction of nucleophile addition to the carbonyl group results in Schiff base formation. In this case, the nucleophile is the amine.

In first step of the mechanism, the amine reacts with the aldehyde or ketone to offer a volatile addition compound known as carbinolamine. In next step, carbinolamine loses water by both acid and base catalyzed pathways. Since carbinolamine is an alcohol, it undergoes acid-catalyzed dehydration which is shown in **Figure 3**.

The Schiff base formation is a series of kinds of reactions that are addition followed by elimination. The chelating cap potential of Schiff bases makes them an intriguing ligand in coordination chemistry combined with their ease of use and flexibility in a variety of chemical environments around the >C=O group [55]. In addition to being useful in the formation of metallic complexes, the formation of Schiff bases is beneficial in a variety of chemical reactions and has essential medical value [56].

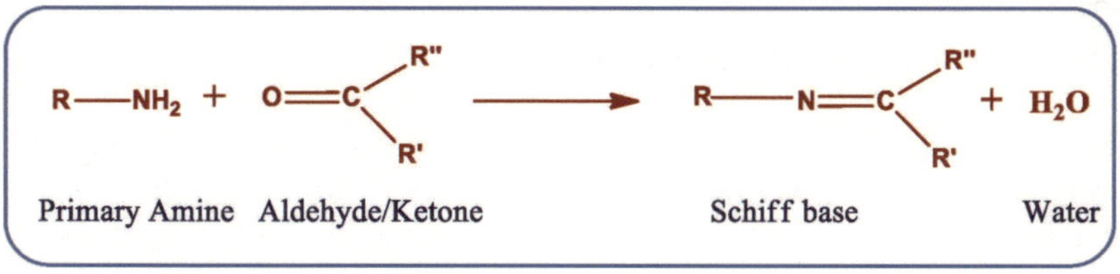

Figure 1.
General synthesis of Schiff base.

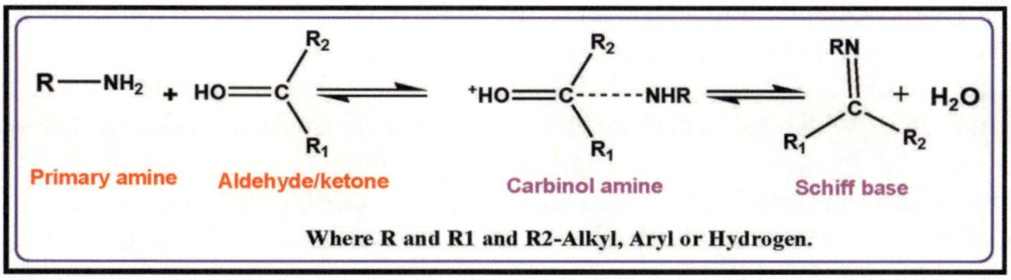

Figure 2.
Formation of Schiff bases via carbinol formation.

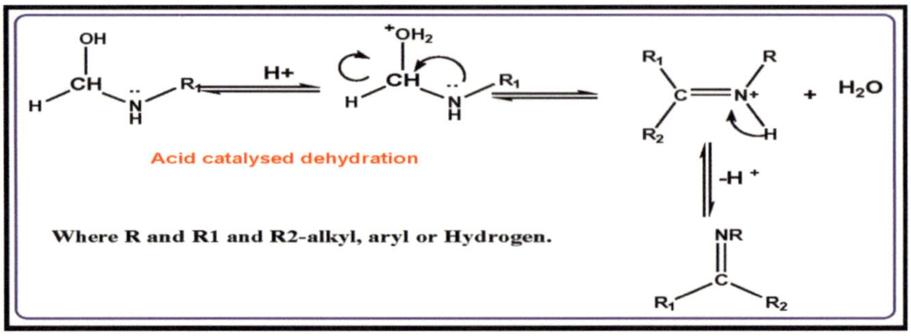

Figure 3.
Formation of Schiff base in acidic medium.

2.1 Chemical properties of Schiff bases

Coordination chemistry has evolved significantly with the help of Schiff bases. Schiff base metal complexes have attracted a lot of interest because of their unique physical and chemical characteristics and numerous applications in a range of scientific fields [57]. Additionally, the heterocyclic compounds provide the fundamental blocks for the synthesis of various Schiff bases. Since Schiff bases physical and biological characteristics have a close relationship to the equilibrium of proton transfer and hydrogen bonding in intermolecular systems, they are of greater importance [58].

Inducing substrate chirality, modifying metal-centered electronic parameters, and enhancing the solubility and stability of homogenous or heterogeneous catalysts are all made feasible by Schiff bases [59]. Schiff base's catalytic activity is demonstrated by several homogeneous and heterogeneous reaction complexes. Transition metal complexes of Schiff bases are involved in multiple biological reactions that are essential to the life process [60]. The metals are necessary for the verification and functioning of biological macromolecules since they can coordinate with the O or N terminals of proteins in various kinds of models [61].

2.2 Synthesis of Schiff base transition metal complexes

Metals and their compounds are required in almost every factor of daily life. Metals are important elements in numerous organic key functions. These factors are beneficial in unique approaches in technology for revolutionizing the manner of communication, nuclear reactor, renewable energy, and generally in drug synthesis.

Depending on their binding nature, metal ions are divided into two classes with the elements like N, O, F, etc. Thus, metallic ions can be identified as hard Lewis acids and soft Lewis acids. Some of the metal ions having d electrons can be donated to suitable ligands to form π-bonds, and a number of the metal ions having vacant d-orbitals can accept electrons from the ligands to form the metal complexes. In the current chemistry, metal complexes are also recognized as coordination complex. The metallic complex is the molecules, which consists of a central metal atom or ion surrounded by different ions, molecules, or atoms that are recognized as ligands or complexing agents [62].

The transition metal due to their vacant d orbital participated in the complex formation. Alfred Werner published the theory of coordination chemistry in 1893. He described the two possibilities in terms of location in the coordination sphere of the metal ions. He also explained the outer and inner sphere complexes with respect to ligands.

Generally, under suitable experimental conditions, metal salts react with Schiff base ligands to form metal complexes [63–65]. On the contrary, the Schiff base metal complexes are made in situ for specific catalytic applications. Generally, equimolar quantity of Schiff bases and metal salt solution were required; however, the amount of metal salt solution varied depending on the type of ligand, such as monodentate, tridentate, or tetradentate. Schiff base metal complexes can be synthesized by using various techniques, such as reflux, heating, microwave, grinding, sonication, and solvents used like ethanol, methanol, or ethyl acetate. Solvent-free synthesis has also been accomplished recently.

Transition metals exhibit different oxidation states and interact with numerous negatively charged molecules, and as a result, transition metal complexes are cationic or anionic species [66, 67]. Schiff base ligands containing azomethine group such a ligand are also known as privileged ligands [68]. Mono or poly nuclear complexes of majority of multidentate N–O, N–N, O–N–S, donor Schiff bases, have important applications.

Schiff base metal complexes have been studied in different areas because of their versatile chemical properties like electrical conductivity, magnetic, catalytic, redox, etc. [69] and wide range of usage in medicinal, chemical, and industrial areas [70]. A number of metals like transition metals and some of the F block elements are also used to synthesize complexes. Complex formation reactions are completely based on nature of ligand and reaction conditions. Complex formation reaction is also very useful to determine quantity of heavy metals that are poisonous to living things.

Generally, a Schiff base is mixed with readily available metallic complexes such as metal amides, alkyls, acetates, or halides to yield metallic complexes. For catalytic applications, this process is appropriate and simple [71].

3. Applications of metal complexes

Schiff base metal complexes have been studied in various chemical and biological areas due to their diverse applications in chemical, pharmacological, dyes-paints, medicinal, material sciences, and industrial areas. The variety of metal complexes is used as a catalyst in many essential chemical reactions like organometallic, metathesis reaction, Suzuki coupling, organic reduction reaction, etc. Schiff bases also show pharmacological activities such as antitubercular, antibacterial, antifungal, antimicrobial, antiviral, anticancer, and wide range of organic activities and commercial applications. Schiff base metal complexes have been of a dominant nature for a long time due to their ability to attach oxygen to redox systems exerting their ability to damage. It also shows the antibacterial activity because of the free radical scavenging property.

Because of the ability of metal complexes to provide synthetic models for metal-containing molecules in metalloproteinase and metalloindole, which are widely employed in the manufacture of fragrances, dyes, agrochemicals, and pharmaceuticals, they have drawn attention in the field of bioinorganic chemistry. Furthermore, preparing metal complexes linked to both synthetic and natural oxygen carriers requires a greater understanding of tetra dentate and tridentate Schiff base complexes. Bioinorganic substances are defined as copper complexes with dimeric units, namely those of the tridentate Schiff base type.

The ability of metallodrugs to cause DNA cleavage is one of the most important criteria for the advancement of metallodrugs as chemotherapeutic agents. Recently, a big number of transition metal complexes have been observed to promote DNA cleavage due to their redox properties. Co (II), Fe (III), and Ru (III) complexes of Schiff bases derived from hydroxyl benzaldehyde are used for the oxidation of cyclohexane into cyclohexanol and cyclohexanone in presence of hydrogen peroxide.

The greenest catalysts are the Fe (III) complexes which are uncommon because the cobalt (II) complexes have high interest in alkane oxidation reactions. New Mn (II) and Mn (III) complexes of substituted N, N′- bis (salicylidine)-1,2- diimino-2-methylene appear to be efficient models for peroxidase activity. New Copper (II) complexes of indoxyl thio-semicarbazone (ITSC) show one pair of well-defined reduction peaks at different potential in the forward scan, which represent the reduction of copper.

3.1 Catalyst

Excellent catalytic activity is observed by several Schiff base complexes in a variety of reactions, including oxygenation, hydrolysis, electro-reduction, and decomposition. Alkene oxygenation is catalyzed by Co (II) aromatic Schiff base four coordinated chelate complexes. Certain Cu complexes increase the rate of hydrolysis obtained from amino acids by 10–50 times compared to the simple copper (II) ion. The iron (II) Schiff base complex is catalytically active for oxygen electro-reduction reactions. Certain metal complexes containing a polymer-bound Schiff base show the catalytic activity for oxidation of ascorbic acid and the breakdown of hydrogen peroxide.

Catalytic activity of cyanohydrin cobaltate complexes is observed at high temperature (1000°C) and in the presence of moisture. At low pressure, Schiff base complexes catalyze the carbonylation of alcohols and alkenes to create aryl propionic acid and its esters. Catalysis reactions are more efficient in terms of energy consumption and waste creation, so it is a significant area of study in both academic and industrial research. These reactions involve catalytically active species that lower the activation energy by forming reactive intermediates through the coordination with organic ligands. The renewal of the catalytically active species should lead to the production of the product.

The efficiency of the catalyst can be calculated on the basis of catalytic cycles passed by one molecule of catalyst. The catalyst should form only labile intermediates with the substrate for efficient regeneration. This concept can be realized using transition metal complexes because metal-ligand bonds are generally weaker than covalent bonds. The possibility of switching reversibly between different oxidation states in redox reactions is due to different oxidation state of transition metals with only moderate differences in their oxidation potentials. Thus many transition metals have been applied as catalysts for organic reactions [72].

Iron has not played a dominant role in catalytic processes. The field of organo-iron chemistry began with the discovery of pentacarbonyl iron by Mond and Berthelot in 1891. Another significant event was the 1951 report on ferrocene. Research was done on Reppe synthesis which is based on iron catalysis. The cross-coupling of Grignard reagents with organic halides, catalyzed by iron, was reported by Kochi and associates in 1971. On the other hand, cross-coupling reactions with transition metals like palladium and nickel gained popularity. In recent times, iron complexes have been used in various reactions, showing the importance of this metal in catalysis.

3.2 Dyes

Transition metals are used to prepare the metal complexes, which show different colors due to different oxidation state and d-d transitions of electrons. Chromium azomethine complex and cobalt complex are used as dyes, to give fast color to wool, leathers, food packages, etc. Azo groups containing metal complexes are used for dying cellulose polyester textiles and mass dye poly-fibers. Salicylaldehyde with diamante Schiff base complex with cobalt has excellent light dye properties and it does not degrade. Some Schiff bases act as a tester or chromogenic reagent for determination of Ni or any other metals present in some natural food samples.

3.3 Polymers

The synthesis and manufacture of polymers have become essential in today's chemical world. Amine-terminated liquid natural rubber (ATNR) is created by photochemically breaking down natural rubber in solution with ethylene-diamine. ATNR and glyoxal are combined, to produce poly Schiff base, which increases antiaging properties. Organo cobalt complexes are used to emulsify and co-polymerize vinyl and divinyl monomers. Tridentate Schiff base serves as an initiator in this process. The alkylation of allylic substrates and the reduction of ketones to alcohols were two important uses of Schiff base complexes.

Polymerization of olefins has increased recently due to catalytic activity of Schiff base complexes in the synthesis of commercially important branched and linear polyethylene's. High-temperature ring-opening polymerization of cycloalkenes with transition metals like ruthenium, molybdenum, and tungsten in the presence of agents like R4Sn or RAlCl2 is possible without affecting the molecular weight of the polymers. Schiff base complex catalyzed ring-opening polymerization of cycloalkenes at low temperature is allowed with control over the molecular weight of the polymers without any adverse reactions.

3.4 Chirality of Schiff bases

Catalysts used in Michael addition reaction are chiral Schiff base complexes of salen and binaphthyl. A new catalytic route for annulations reaction using Schiff base complexes is a current area of research, although the hetero annulations reaction has been reported.

The complexes of nickel (II) and copper (II) ions have increased enantioselectivity in the alkylation of enolates. According to this research, Schiff base complexes can act as catalysts to affect the yield and selectivity of chemical transformations. Hence, it is important to evaluate the applications of complexes and determine the appropriate function of Schiff base complexes in different processes.

According to the binding site of ligands with metal ions, it is classified as the mono, di, tri, tetra, and multi-dentate chelating ligands. The synthesis of chiral complexes become an important area of current research in coordination chemistry due to stereoselectivity of metal complexes of chiral Schiff base ligands in organic transformation. The chiral binaphthyl Schiff base ligands were potentially valuable in various metal-mediated catalytic reactions.

Chiral Schiff base ligands with titanium (IV), vanadium (IV), copper (II), or zinc (II), transition metal ions were used in various asymmetric chemical transformations. The addition of trimethylsilyl cyanide to benzaldehyde in the presence of titanium (IV) ions resulted in trimethylsilyl cyanohydrin 40–85% enantioselectivity.

3.5 Biological applications

It is known that ternary complexes are essential for the activation of enzymes as well as the storage and movement of active compounds. In the reaction, these components give a lot of surface area for the reactants to be absorbed and move them closer to one another. Biological activity has been revealed by the ternary and binary transition metal complexes [73]. Biological systems often include transition metal-ligand catalysts. Recently, metal complexes of certain N-/O donor ligands have received a lot of attention, due to their stronger antifungal and antibacterial properties than the parent ligands. Ternary complexes with an amino acid as a secondary ligand are important as they can serve as models for complexes between metal ions and enzymes.

A review of the literature indicated that the synthesis of heterogeneous metal complexes has received less attention. Some metal-ligand catalysts have been found to catalyze reactions such as oxidation, oxidative cleavage, and hydroformylation and decomposition of hydrogen peroxide. Phthalocyanines are widely used in diverse areas. Around the world, many scientists working on creating and developing new metal complexes to explore their potential biological and catalytic uses. Additionally, efforts are taken to synthesize the metal complexes to use in agriculture, medicine, and other industries. The study of various transition metal complexes and metal alkylated catalysts has attracted a lot of attention in recent years. The interest in the design and synthesis of novel transition metal complexes lies in their biological and catalytic activity in numerous reactions.

In chemical fields, there are immense applications of Schiff base alkylated groups and ligands, as catalysts or complexes. Schiff bases with variable donation sites could be monodentate, bidentate, tridentate, or tetradentate, forming mono- or polynuclear complexes. Tetradent Schiff base complexes are employed to design metal complexes related to synthetic and natural oxygen carriers.

3.6 Antiviral activities

Gossypol's Schiff bases and their Ag (I) complexes have strong antiviral activity and inhibition against the Cucumber mosaic virus. In this modern era, producing antiviral medications is a significant task, and our research has overcome those obstacles. The synergistic activity of some tetra dentate Schiff base and its metal complexes with Mn (II), Ni (II), Cu (II), and Zn (II) demonstrate a variety of effects on a membrane in amylose synthesis, as well as therapeutic activities, anticancer, and cytotoxic properties [65]. Amylase transportation through the membrane was enhanced by Zn (II) and Mn (II) complexes, but it was inhibited by Ni (II) and Cu (II) complexes, which also possessed simple harmonic generation activity [74]. In the degradation of aldose-inferred Schiff bases carnosine and serine are effective trans-glycating agents.

3.7 Insecticide, pesticide, and repellent activity

Copper complexes of semicarbazones and phenolic hydrazone have been utilized as herbicides, insect poisons, nematocides, and rodenticides. Phenoxyacetic corrosive and its subordinates are used as herbicides. Aryloxyacetate anions act as corrosive ligands while (carboxylato) copper (II) complex have herbicidal activities. Pesticide exercises of Schiff's bases like; N-(1-phenyl-2-hydroxy-2- phenylethylidine) etc. were synthesized from benzoin, salicylaldehyde, 2-aminophenol and 2, 4-dinitrophenyl hydrazine and their copper complex has been concentrated against Tribolium castaneum.

Complex of copper and N-salicylidene-O, S-dimethyl thiophosphorylimine (HSMa) showed a lot higher pesticide rates. Metal like aluminum, cobalt, and mercury are also used as insecticide and pesticide chemicals. Many natural extracts that can have chemical composition with aromatic compounds were used as an effective repellent for many insect pests. Schiff base containing metal complexes was used as a contact poison.

4. Conclusion

Many chemists around the globe are busy designing and synthesizing novel metal complexes to test their possible biological/catalytic applications. Efforts are also directed toward the synthesis of metal complexes for other industrial, agricultural, and medicinal applications. Recent years have witnessed a great deal of interest in the research of different types of metal alkylated catalysts and metal complexes of transition elements. The interest in the design and synthesis of novel transition metal complexes containing metals lies in their biological and catalytic activity in many reactions. Many metal complexes possess interesting biological properties, such as antibacterial and antitumor activities.

There are numerous applications for the complexes of Schiff base alkylated groups or catalysts made, depending on the ligands and group characteristics. Mono- or polynuclear complexes can be formed by Schiff bases with varying donation sites, which can be bidentate, tridentate, tetradentate, or monodentate. To synthesize metal complexes attached to artificial and natural oxygen carriers, tetradent Schiff base complexes are used.

Schiff-base ligand-based metallic complexes have diverse and interdisciplinary uses, so it is a very important field of study. Specifically, the study of Schiff base complexes in biology and catalysis has demonstrated their tremendous potential as potent antimicrobial and anticancer agents and efficient catalysts for various chemical reactions.

Author details

Shilpa Laxman Sangle[1*], Rekha Barku More[2] and Savita V. Thakare[3]

1 Department of Chemistry, MVP Samaj's Arts, Commerce and Science College, Nandgaon, India

2 Department of Chemistry, MVP Samaj's SSSM Arts, Science and Commerce College, Saikheda, India

3 Department of Chemistry, K.S.K.W. Arts, Commerce and Science College, CIDCO, Nashik, India

*Address all correspondence to: shilpasangle620@gmail.com

References

[1] Schiff H. Mittheilungen aus dem Universitäts-laboratorium in Pisa: 2. Eine neue Reihe organischer Basen [communications from the university laboratory in Pisa: 2. A new series of organic bases]. Annalen der Chemie und Pharmacie (in German). 1864;**131**:118-119. DOI: 10.1002/jlac.18641310113

[2] Gupta K, Sutar A. Coordination Chemistry Reviews. 2008;**252**:1420

[3] Kathryn L, Franz K. Application of metal coordination chemistry to explore and manipulate cell biology. Chemical Review. 2009;**109**:4921-4960

[4] Ommenya FK, Nyawade EA, Andala DM, Kinyua J. Synthesis, characterization and antibacterial activity of Schiff Base, 4-Chloro-2-{(E)-[(4-Fluorophenyl)imino]methyl} phenol metal (II) complexes. Journal of Chemistry (New York). 2020;**2020**. DOI: 10.1155/2020/1745236

[5] Majumdar D et al. Coordination of N, O-donor appended Schiff base ligand (H 2 L1) towards zinc (II) in presence of pseudohalides : Syntheses, crystal structures, photoluminescence, antimicrobial activities and Hirshfeld surfaces. Journal of Molecular Structure. 2018;**1155**:745-757

[6] Borase JN, Mahale RG, Rajput SS, Shirsath DS. Design, synthesis and biological evaluation of heterocyclic methyl substituted pyridine Schiff base transition metal complexes. SN Applied Sciences. 2021;**3**:1-13

[7] Shetty P. CrossRef Citations to Date 0 Altmetric Review Articles Schiff Bases: An Overview of their Corrosion Inhibition Activity in Acid Media against Mild Steel 2019. Taylor Fr; 2018. pp. 985-1029

[8] Chełmieniecka A, S, Rzuchowska A, Szczupak AM, Schilf W, Rozwadowski Z. New one pot Pd (II) and Zn (II) complexes of Schiff bases, derivatives of 1 amino 1 deoxy d sorbitol: Spectroscopic studies and biological and catalytic activities. Applied Organometallic Chemistry. 2020;**34**:1-8

[9] Moffett RB, Rabjohn N. Organic Synthesis. New York: John Wiley Sons; 1963. pp. 605-608

[10] Vats V, Upadhyay RK, Sharma P. Synthesis and antifungal activities of 2-ketophenyl-3-substituted. International Journal of Theoretical & Applied Sciences. 2009;**1**:24-26

[11] Taguchi K, Westheimer FH. Catalysis by molecular sieves in the preparation of Ketimines and Enaminestle. The Journal of Organic Chemistry. 1971;**36**:1570-1572

[12] Ren BELJ. Synthesis of sterically hindered imine. The Journal of Organic Chemistry. 1993;**58**:5556-5559

[13] Uddin MN, Chowdhury DA, Rony M. Complexes of Schiff bases derived from synthesis, characterization and antibacterial screening. American Journal of Chemistry and Applications. 2014;**1**:12-18

[14] Parikh KS, Vyas SP, Parikh KS, Appl A, Res S. Design and spectral analysis of novel Schiff base derived with acetophenone derivatives. Archives of Applied Science Research. 2012;**4**:1564-1566

[15] Todd D. Schiff Base puzzle project. Journal of Chemical Education. 1992;**69**:584-588

[16] Naeimi H, Rabiei K. Convenient, mild and one-pot synthesis of double Schiff bases from three component

reaction of salicylaldehyde, ammonium acetate and aliphatic aldehydes accelerated by NEt3 as a base. Journal of the Chinese Chemical Society. 2007;**54**:1293-1298

[17] Abdalrazaq EA, Al-Ramadane OM, Al-Numa KS. Synthesis and characterization of dinuclear metal complexes stabilized by tetradentate schiff base ligands. American Journal of Applied Sciences. 2010;**7**:628-633

[18] Reddy PS, Ananthalakshmi PV, Jayatyagaraju V. Synthesis and structural studies of first row transition metal complexes with tetradentate ONNO donor schiff base derived from 5-acetyl 2,4-dihydroxyacetophenone and ethylenediamine. E-Journal of Chemistry. 2011;**8**:415-420

[19] Ustynyuk YA. Metal free methods in the synthesis of macrocyclic Schiff bases. Chemical Reviews. 2007;**107**:683-689

[20] Pandeya SN, Rajput N. Synthesis and anticonvulsant activity of various Mannich and schiff bases of 1,5-benzodiazepines. Der Pharmacia Lettre. 2012;**4**:938-946

[21] Selwin Joseyphus R, Sivasankaran Nair M. Synthesis, characterization and biological studies of some Co(II), Ni(II) and Cu(II) complexes derived from indole-3-carboxaldehyde and glycylglycine as Schiff base ligand. Arabian Journal of Chemistry. 2010;**3**:195-204

[22] Ndahi NP, Synthesis YNPUKS. Characterization & antibacterial studies of some Schiff Base complexes of cobalt(II), nickel(II) and zinc(II). Asian Journal of Biochemical and Pharmaceutical Research. 2012;**2**:308-316

[23] Brodowska K, Łodyga-Chruścińska E. Schiff bases—Interesting range of applications in various fields of science. Chemik. 2014;**68**:129-134

[24] Rashmi A, Ashish K, Gill NS, Rana AC. Synthesis of novel Schiff bases of 5-aminosalicylic acid by grinding technique and its evaluation for anti-inflammatory and analgesic activities. International Research Journal of Pharmacy. 2012;**3**:185-188

[25] Clercq BD, Lefebvre F. Immobilization of multifunctional Schiff base containing ruthenium complexes on MCM-41. Applied Catalysis A2. 2003;**47**:345-355

[26] Wadher SJ, Puranik MP, Karande NA, Yeole PG. Synthesis and biological evaluation of Schiff Base of Dapsone and their derivative as antimicrobial agents. International Journal of PharmTech Research. 2009;**1**:22-33

[27] Kumar S, Dhar DN, Saxena PN. Applications of metal complexes of Schiff bases-A review. Journal of Scientific and Industrial Research. 2009;**68**:181-187

[28] Lin HC, Huang CC. Synthesis of alkynylated photo-luminescent Zn(II) and Mg(II) Schiff base complexes. Chemical Sciences Journal. 2007;**7**:781-785

[29] Basak S, Sen S, Banerjee S. Three new pseudohalide bridged dinuclear Zn (II) Schiff base complexes: Synthesis, crystal structures and fluorescence studies. Polyhedron. 2007;**26**:5104-5111

[30] Rashmi A, Gills KA, Rana NS. Synthesis of novel Schiff base of 5-aminosalicylic acid by grinding techniques and its evaluation for anti-inflammatory and analgesic activities. International Research Journal of Pharmacy. 2012;**3**:112-119

[31] Lestari WW, Lönnecke P, Streit HC, Schleife F, Wickleder C, Hey-Hawkins E. A chiral two-dimensional coordination polymer based on CuII and (S)-4,4′-bis(4-carboxyphenyl)-2,2′-bis(diphenylphosphinoyl)-1,1′-binaphthyl: Synthesis, structure, and magnetic and optical properties. Inorganica Chimica Acta. 2014;**421**:392-398. DOI: 10.1016/j.ica.2014.06.006

[32] Swiegers GF, Malefetse TJ. Classification of coordination polygons and polyhedra according to their mode of self-assembly. 2. Review of the Literature. 2002;**225**:91-121

[33] Jin HY, Chen JZ, Wang XW, Hou SE. Solvothermal synthesis and crystal structure of a 18-membered macrocycle Schiff base dinuclear copper(II) complex: $Cu_2(NO_3)_4$(APTY)4 (APTY = 1,5-dimethyl-2-phenyl-4-{[(1E)-pyridine-4-ylmethylene]amino}-1,2-dihydro-3H-pyrazol-3-one). Journal of Chemical Crystallography. 2009;**39**:182-185

[34] Gryca I, Machura B, Shul'pina LS, Shul'pin GB. Synthesis, structures and catalytic activity of p-tolylimido rhenium(V) complexes incorporating quinoline-derived ligands. Inorganica Chimica Acta. 2017;**455**:683-695

[35] Critelli RAJ, Sumodjo PTA, Bertotti M, Torresi RM. Influence of glycine on Co electrodeposition: IR spectroscopy and near-surface pH investigations. Electrochimica Acta. 2018;**260**:762-771

[36] Kelode SR, Mandlik PR, Aswar AS. Synthesis of hydrazones schiff bases and microbiological evaluation of lsonicotinoyl hydrazide with different acetophenone. Oriental Journal of Chemistry. 2011;**27**:1053-1062

[37] Abu-Khadra AS, Farag RS, Abdel-Hady AE-DM. Synthesis, characterization and antimicrobial activity of Schiff Base (E)-N-(4-(2-hydroxy benzylideneamino) phenylsulfonyl) acetamide metal complexes. American Journal of Analytical Chemistry. 2016;**07**:233-245

[38] Mahmudov KT, Kopylovich MN, Guedes MFC, Pombeiro AJL. Non-covalent interactions in the synthesis of coordination compounds: Recent advances. Coordination Chemistry Reviews. 2016:1-19. DOI: 10.1016/j.ccr.2016.09.002

[39] Winter A, Hager MD, Newkome GR, Schubert US. The marriage of terpyridines and inorganic nanoparticles: Synthetic aspects, characterization techniques, and potential applications. Advanced Materials. 2011;**23**(45):5728-5748. DOI: 10.1002/adma.201103612

[40] Kar AK et al. Ligand substitution and electron transfer reactions. RSC Advances. 2014;**4**:58867-58879. DOI: 10.1039/c4ra10324f

[41] Dhanwe VP, Joshi PG, Kshirsagar AS, Dhapte VV, Khanna PK. Synthesis, characterization and nanochemistry of novel antibacterial copper (II) semicarbazone complexes. ChemistrySelect. 2018;**3**:7548-7560

[42] Macı B, Castin A. Copper complexes with sulfonamides: Crystal structure and interaction with pUC18 plasmid and hydrogen peroxide. Inorganica Chimica Acta. 2003;**342**:241-246

[43] Transition B, Werner A. Investigating complex ions of copper(II) background. pp. 198-204

[44] Topală T, Bodoki A, Oprean L, Oprean R. Experimental techniques employed in the study of metal complexes-DNA-interactions. Farmácia. 2014;**62**:1049-1061

[45] Paradella TC, Bottino MA. Scanning electron microscopy in modern dentistry research. Brazilian Dental Science. 2012;**15**:43-48

[46] Khandar AA, Hosseini-Yazdi SA, Zarei SA. Synthesis, characterization and X-ray crystal structures of copper(II) and nickel(II) complexes with potentially hexadentate Schiff base ligands. Inorganica Chimica Acta. 2005;**358**:3211-3217

[47] Osman Dolu MC, Issahhaya BANS. A novel and synthetically facile coumarin-thiophene-derived Schiff base for selective fluorescent detection of cyanide anions in aqueous solution: Synthesis, anion interactions, theoretical study and DNA-binding properties. Tetrahedron. 2018;**74**:6897-6906

[48] El-Faham A et al. Ultrasonic promoted synthesis of novel s-triazine-Schiff base derivatives; molecular structure, spectroscopic studies and their preliminary anti-proliferative activities. Journal of Molecular Structure. 2016;**1125**:121-135

[49] Yusenko KV, Sukhikh AS, Kraus W, Gromilov SA. Synthesis and crystal chemistry of octahedral rhodium(III) chloroamines. Molecules. 2020;**25**:76. DOI: 10.3390/molecules25040768

[50] Raman N, Ravichandran S, Thangaraja C. Copper (II), cobalt (II), nickel (II) and zinc (II) complexes of Schiff base derived from benzil-2, 4-dinitrophenylhydrazone with aniline. Journal of Chemical Sciences. 2004;**116**:215-219

[51] Raman M, Muthuraj V. Synthesis charecterisation and electrochemical behaviour of Cu, CO, Ni, and Zn complexes derived from acetyl acetone and p anisidineand thire antimicrobial applications. Journal of Chemical Sciences. 2003;**115**:161-167

[52] Mermer A et al. Synthesis of novel Schiff bases using green chemistry techniques; antimicrobial, antioxidant, antiurease activity screening and molecular docking studies. Journal of Molecular Structure. 2019;**1181**:412-422

[53] Break M, Bin K, Tahir MIM, Crouse KA, Khoo TJ. Synthesis, characterization, and bioactivity of schiff bases and their Cd2+, Zn2+, Cu2+, and Ni2+ complexes derived from chloroacetophenone isomers with S-benzyldithiocarbazate and the X-ray crystal structure of S-benzyl-β-N-(4-chlorophenyl) methylenedi. Bioinorganic Chemistry and Applications. 2013

[54] Rublein K. The reaction of benzaldehyde and aniline by Kurt Rublein. Molecules. 1998;**2**:1-8

[55] Gavhane VS, Rajbhoj AS, Gaikwad ST. X-ray diffraction study and biological analysis of transition metal complexes of N-4-disubstituted thiosemicarbazone. Research Journal of Chemical Sciences. 2015;**5**:33-37

[56] Nwabueze JN, Salami HA. Complexes of 4-Cyanobenzal dehydeisonicotinic acid hydrazone with some transition M(II) sulphates. International Journal of Research Inorganic Chemistry. 2013;**2**:1-3

[57] Katwal R, Kaur H, Kapur BK. Applications of copper—Schiff's base complexes: A review. Scientific Reviews and Chemical Communications. 2013;**3**:1-15

[58] Tomma JH, Khazaal MS, Al-Dujaili AH. Synthesis and characterization of novel Schiff bases containing pyrimidine unit. Arabian Journal of Chemistry. 2014;7:157-163

[59] Khalil BI. Study of suspended solution from blending three polymer metal complexes as antimicrobial. Chemical and Process Engineering Research. 2014;**21**:1-8

[60] Martins LMDRS, Hazra S, Guedes Da Silva MFC, Pombeiro AJL. A sulfonated Schiff base dimethyltin(iv) coordination

polymer: Synthesis, characterization and application as a catalyst for ultrasound- or microwave-assisted Baeyer-Villiger oxidation under solvent-free conditions. RSC Advances. 2016;**6**:78225-78233

[61] Prakash A, Adhikari D. Application of Schiff bases and their metal complexes-A review. International Journal of ChemTech Research. 2011;**3**:1891-1896

[62] Popkov A, Gee A. Chiral nucleophilic glycine and alanine synthons: Nickel (II) complexes of Schiff bases or alanine. Transition Metal Chemistry. 2002;**2**:884-887

[63] Tolstikov AH, Khlebnikova TB, Tolstikov HA. Chiral organophosphorus ligands. Synthesis and application in asymmetric metal-complex catalysis. Chemistry and Computational Simulation. Butlerov Communications. 2002;**2**:27-54

[64] Park K. Using Phosphine Aldehydes to Generate New Transition Metal Complexes and the Synthesis of Chiral NHC-Amino Ligands by Using Phosphine Aldehydes to Generate New Metal Complexes and the Synthesis of Chiral NHC-Amino Ligands. 2013. Available from: https://hdl.handle.net/1807/35134

[65] Hasaninejad A, Zare A, Mohammadizadeh MR, Shekouhya M. Oxalic acid as an efficient, cheap, and reusable catalyst for the preparation of quinoxalines via condensation of 1,2-diamines with α-diketones at room temperature. Arkivoc. 2008;**13**:28-35

[66] Ahmed Riyadh M, Yousif Enaam I, Hasan Hasan A. Metal complexes of macrocyclic Schiff-base ligand: Preparation, characterization, and biological activity. The Scientific World Journal. 2013;**23**:1-7

[67] Mishra AP, Sharma N, Jain RK. Microwave synthesis, spectral, thermal and antimicrobial studies of some Ni(II) and Cu(II) Schiff Base complexes. Open Journal of Synthesis Theory and Applications. 2013;**02**:56-62

[68] Da Silva CM et al. Schiff bases: A short review of their antimicrobial activities. Journal of Advanced Research. 2011;**2**:1-8

[69] Kajal A, Bala S, Kamboj S, Sharma N, Saini V. Schiff Bases: A versatile pharmacophore. Journal of Catalysts. 2013. DOI: 10.1155/2013/893512

[70] Jesmin M, Ali MM, Khanam JA. Antitumour activities of some schiff bases derived from benzoin, salicylaldehyde, amino phenol and 2,4 dinitrophenyl hydrazine. Thai Journal of Pharmaceutical Sciences. 2010;**34**:20-31

[71] Clark MA, Hilderbrand SA, Lippard SJ. Synthesis of fluorescein derivatives containing metal-coordinating heterocycles. Tetrahedron Letters. 2004;**45**:7129-7131

[72] Ali MM, Jesmin M, Salam SMA, Khanam JA, Islam MF. Pesticidal activities of some Schiff bases derived from benzoin, salicylaldehyde, aminophenol and 2, 4 dinitrophenyl hydrazine. Journal of Scientifc & Research. 2009;**1**:641-646

[73] Ishii T, Matsuzawa H, Vairappan CS. Repellent activity of common spices against the rice weevil, Sitophilus zeamais Motsch (Coleoptera, Curculionidae). Journal of Tropical Biology & Conservation. 2010;**7**:75-80

[74] Kundu A et al. Microwave assisted solvent-free synthesis and biological activities of novel imines (Schiff bases). Journal of Environmental Science and Health, Part B Pesticides Food Contaminants, and Agricultural Wastes. 2009;**44**:428-434

Chapter 6

Crystal Structure Prediction for Powder XRD of Polymorph toward Intractable Metal Complexes

Yuya Higashi, Rin Tsuchiya, Diasuke Nakane and Takashiro Akitsu

Abstract

Originating from the challenges of using ordinary methods for determining the structure of salen-type nickel(II) complexes and the limitations of crystallography of unknown organic powders, which has not been reported for many years, this chapter mainly reviews crystal structure prediction, polymorphism, phase transitions, and the application of computational methods. It also states that new research is incorporating computational chemistry for analyzing powder XRD. In particular, we will explore the possibility of predicting polymorphic crystal structures, which are of academic and industrial interest and applying them to metal complexes and organic-inorganic hybrid materials. There are also high hopes for the recent development of electron diffraction.

Keywords: salen-type nickel(II) complex, X-ray diffraction, crystal structure prediction, electron diffraction, polymorphism, phase transition

1. Introduction

Powder X-ray structural analysis is sometimes used when conventional single-crystal X-ray structural analysis is difficult. However, analysis of powder requires "favorable conditions" to be met, and it is not uncommon for the analysis to fail. Crystal polymorphism and crystal solvents are an inherent difficulty. The focus here will be on the crystal structure of compounds based on powder analysis, which can potentially take on different crystal structures depending on temperature and solvent conditions.

Several well-known examples of inorganic compounds exhibiting structural phase transitions and crystal polymorphisms are known. For example, ZnS (sphalerite and wurtzite), TiO_2 (anatase and rutile), and $BaTiO_3$ (rhombohedral-orthorhombic-tetragonal [dielectric]-cubic) are widely known. As we learn in basic chemistry, ionic bonding is based on low directionality, and ions that can approximate a spherical shape can differ in packing. This is due to the difference in ion arrangement.

On the other hand, many examples of organic compounds exhibiting structural phase transitions and crystal polymorphisms are also known. In general, organic compounds have more "flexibility" in their three-dimensional structures, and intermolecular interactions than inorganic compounds, so structural phase transitions and crystal polymorphisms that accompany changes and differences in molecular structure are common. In addition, although not polymorphic, the same molecule can have different crystal structures depending on the presence or absence of a crystal solvent (often water), which is an important aspect of drug development. Now, in the case of a metal complex in which a metal ion and an organic ligand are combined, it is necessary to consider both elements (although it can be treated in the same way as an organic compound in many cases).

Structural phase transitions and crystal polymorphisms of metal complexes have recently been reported by Miyamura and Honda [1, 2]. Salen-type nickel(II) complexes with long-chain alkyl groups exhibit the "cold-crystallization" phenomenon, which is often seen in conventional high-molecular compounds, even though they are low-molecular-weight compounds. Here, we do not refer to the "anomalous thermal behavior" of "cold-crystallization" related to structural phase transitions and crystal polymorphisms. In terms of structural chemistry, conformational changes in long-chain alkyl groups and accompanying crystal structural changes are important. Alkyl chains tend to assemble parallel to each other as a result of van der Waals interactions, and this interaction affects the crystal structure. Therefore, the structure of alkyl derivatives depends on the chain length. Therefore, alkylation of rigid molecules can significantly change their physical properties, such as thermal properties, crystal structure, and the structure of adsorbed monolayers. However, although there are many examples of salen-type metal complexes, the fact that this phenomenon was observed in nickel(II) complexes requires consideration of special circumstances. In fact, salen-type ligands can be distorted into the coordination structure oriented by metal ions while maintaining planarity [3, 4]. Furthermore, it is known that salen-type nickel(II) complexes tend to form a dimer structure in the crystal. The structure is built up in a centrosymmetric dimer characterized by a weak but significant Ni-Ni intermolecular interaction with a Ni-Ni distance of about 3.16–3.20 Å.

Upon successful recrystallization, microscopic observation of the crystal quality revealed two different crystal habits depending on the solvent used. Crystals grown from acetone, ethyl acetate, toluene, and N-methyl-2-pyrrolidone ("NMP") were plate-like, while those grown from chloroform, dichloromethane ("DCM"), dimethyl-formamide ("DMF"), acetic acid, and methanol were needle-like. Crystals grown from ethanol were either plate-like or very fine needle-like. X-ray crystallography revealed that the plate-like crystals always corresponded to the solvent-free compound, and the needle-like crystals to the five solvates of Ni(salen): Ni(salen)-chloroform, Ni(salen)-DCM, Ni(salen)-DMF, Ni(salen)-AcOH, and Ni(salen)-1.5MeOH. Crystals of the ethanol solvate were not successfully isolated.

When crystals were grown by evaporation from ethanol solutions near room temperature, they were very fine and needle-like, but their quality was not sufficient for X-ray crystal structure determination. All successful recrystallizations took several hours to several days. Recrystallization from pure dimethyl sulfoxide (hereafter DMSO) or pure ethylene glycol still did not yield crystals; recrystallization from water failed because Ni(salen) is insoluble in water; recrystallization from a mixture of DMSO and water (9:1) yielded crystals of solvent-free 1. No attempt was made to recrystallize Ni(salen) from other mixed solvents.

The nature of the weak but important metal-metal interactions observed in square-plane d^8 complexes has been explained by the concept of overlapping metal-metal orbitals, i.e., the overlap of the p_z (empty) and d_{z2} (fully occupied) orbitals of one d^8 atom with those of the neighboring d^8 atom. The intermetallic distance is then governed by the amount of overlap between metal-metal orbitals. In the reported structures of solvent-free compounds and their solvates, neighboring Ni(salen) molecules always exhibit weak intermetallic interactions, which can be found along centrally symmetric dimers or one-dimensional chains. In this series of compounds, the metal-metal interactions found in the dimer units appear to be slightly stronger than those found in the one-dimensional chains, but the intermetallic distances in the former group differ significantly [Δ(Ni---Ni) max-min ≈ 0.26 Å]. However, there is no apparent correlation between the strength of the metal-metal interactions and the nature of the intermolecular solvent-salen interactions; further investigation of the metal-metal interactions found in 1 should be done in a separate paper. The reason for the preferential formation of either dimeric or one-dimensional chain structures in the seven structures of Ni(salen) and its solvates is still unclear. While dimeric units seem to be preferred in these structures, one solvated structure contains one-dimensional chains, which are linked by solvent molecules via the motif Ni(salen)---solvent---solvent. Such a motif may provide extra rigidity to the monomer units and favor the construction of one-dimensional chains.

There are also reports of the crystal structures of salen-type nickel(II) complexes with some simple substituents such as ethylenediamine, 1,2-diaminocyclohexane, 1,2-diphenylethylendiamine [5–7]. Therefore, not only the characteristic dimer structure but also from the powder X-ray diffraction pattern applicable to structural phase transitions and crystal polymorphisms, or in combination with other related techniques, (racemic) propylenediamine derivatives salen-type (**Figure 1**). An attempt was made to analyze the crystal structure of nickel(II) complexes.

The first difficulty was that single crystals of sufficient size could not be grown, although single crystals of small size could be obtained by X-ray diffraction. Therefore, the in-house apparatus could not obtain a decent diffraction peak.

The next difficulty was that the structural analysis of unknown organic crystals from synchrotron radiation powder X-ray diffraction patterns failed at the stage of lattice parameter indexing. Taking advantage of simple substituents, we were preparing for Rietveld analysis and fitting using the optimized structure by density functional theory (DFT) calculation as a model, but we did not proceed to that stage. Synchrotron radiation X-ray diffraction was able to be measured on a very small single crystal sample and a randomly oriented fine powder sample (**Figure 2**).

A final difficulty arose when electron diffraction was attempted in order to perform diffraction data measurements and structural analysis on “single crystals” that were too small. Unexpectedly, a weak regular diffraction pattern was observed for a specific one-dimensional (probably c-axis) direction, and no diffraction pattern was observed for other directions (**Figure 3**). This suggests that although there is barely regular molecular alignment in one direction (stacking of planar ligands due to the formation of the dimer structure typical of nickel(II) complexes), there is order in

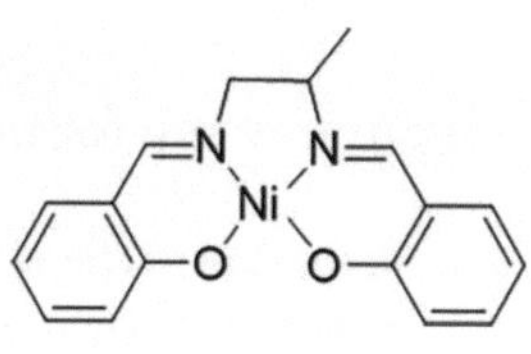

Figure 1.
Chemical structure of the racemic nickel(II) complex.

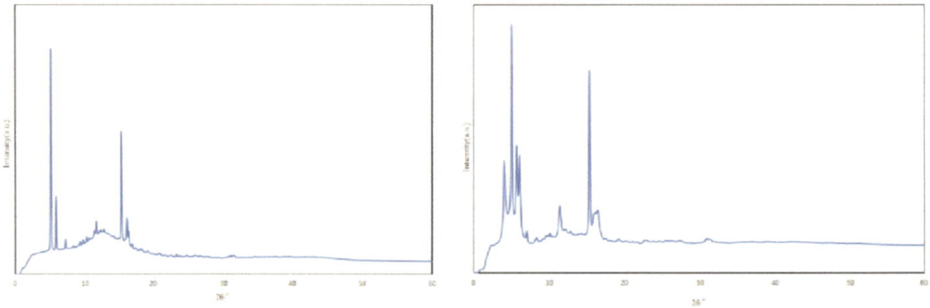

(a) (b)

Figure 2.
Powder XRD patterns of the nickel(II) complexes as (a) powder and (b) single crystal.

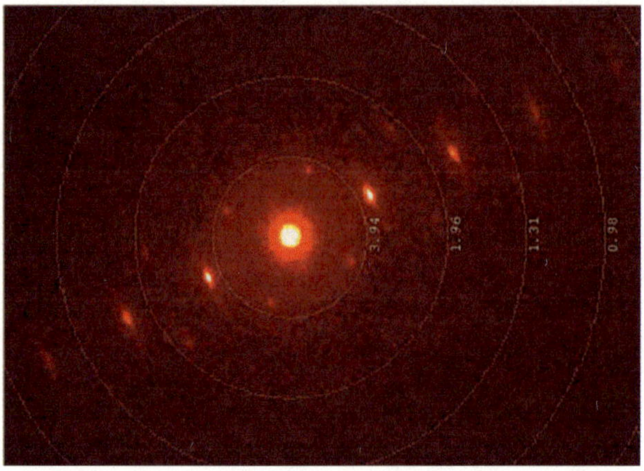

Figure 3.
Electron diffraction patterns (along the c-axis) of the nickel(II) complex.

the direction perpendicular to it. It was speculated that the crystal packing is weakly regular and unique. Under these circumstances, we have come to believe that there have been no reports of single-crystal X-ray structural analysis of nickel(II) ion-containing molecules, which are the simplest class of molecules.

Based on these backgrounds and motives, are there any research examples or findings that can provide hints for crystal structure analysis of such difficult crystal polymorphs (structural phase transitions) by X-ray powder diffraction? Although there are many organic compounds, we decided to investigate research examples of structure analysis by powder X-ray diffraction or three-dimensional electron diffraction while predicting the crystal structure of crystal polymorphs.

2. Crystal structure prediction polymorph for powder XRD

In addition to cases where it is difficult due to the reasons mentioned above, a number of crystal structures have been reported by powder analysis of compounds that can have different crystal structures. Particularly in the case of organic compounds for applied purposes, strong motivation and great significance are the driving force. However, in general, analysis is often not easy and is done on a case-by-case basis. Here, the characteristics of compounds, analysis methods, and additional methods for analysis (other measurements and computational chemistry) will be reviewed.

Watt et al. [8] reported the crystal polymorphism of 5-aminouracil derivatives. In order to obtain polymorphic single-crystal samples, sublimation or methods using various solvents were employed, and data with different crystal shapes and XRD patterns were obtained. A comparison of the molecular structures of the polymorphs revealed different hydrogen bonding patterns and differences in the tetrahedral conformation of the proton-donor amino groups. The difference in lattice energies is mentioned. They are discussing the crystal structure based on *synchrotron radiation X-ray diffraction data from single crystals*. However, the powder XRD pattern does not give uniform and clean data due to the influence of "phase impurities." Indeed, the nucleobase derivative 5-aminouracil (AUr) is intriguing for its biological activity, but the solid-state structure of this compound remains elusive because of its tendency to crystallize as an aggregate of microcrystalline particles. Both crystallization of AUr from solution and crystallization by sublimation were revisited, and large, albeit difficult, single crystals of AUr were grown. Synchrotron X-ray diffraction of the single crystal revealed the structure of this compound for the first time. The structure obtained was one of the previously calculated structures, but the corresponding calculated structure was more than 10 kJ mol^{-1} above the minimum and was not considered experimentally feasible (early efforts to predict crystal structures had assumed that AUr was rigid in an optimized conformation of isolated molecules). Periodic *electron-level modeling* (PBE-TS optimization) revisited these crystal structures and yielded more realistic relative lattice energies, but it required the use of experimental cell parameters in good agreement with the experimental powder model. Powder X-ray diffraction of the precipitated material and mirror image *Raman analysis* of individual crystal aggregates indicated the possible presence of at least one other polymorph in small amounts, which is consistent with the thermodynamic validity of the *computer-generated polymorphs*. This experimental polymorph was isolated from multiple crystal growth methods and identified as one of the computationally generated crystal structures. Although only one polymorph of this compound was structurally characterized, there is evidence that additional polymorphs are likely present in the bulk growth product. This structural information provides useful insight into the effect of 5-position substitution on uracil derivatives. This study illustrates the utility of inexpensive crystal structure prediction (CSP) studies using rigid molecules and the challenges faced by current computational modeling of electronic and geometric changes associated with -NH_2 group delocalization and thermal expansion in the organic solid state. Nevertheless, the energetic competitiveness of the various crystal structures provides evidence that AUr molecules offer a variety of hydrogen bonding interactions and are important nucleobase derivatives in novel oligonucleotide assemblies.

Gushurst et al. [9] discovered a new polymorph of the organic compound ROY, for which 11 polymorphs have already been discovered. In this study, the structure was analyzed by *Rietveld analysis of powder X-ray diffraction as a matching polymorph in order to match the previously reported experimental data and crystal structure prediction.* A novel polymorph of ROY was discovered using a thermal method. The crystal structure was elucidated by matching experimental data with powder X-ray diffraction patterns calculated from previously published crystal structure predictions. The structure was further refined by the Rietveld method. This new polymorph was named PO13 because of its pumpkin-orange color. At room temperature, PO13 is not as stable as Form Y. After 5 years, a sample of 99% pure PO13 was converted to a mixture of 45% PO13 and 55% Form Y. A new polymorph of ROY was discovered in 2013. Recently, the structure was refined using the Rietveld method on the PXRD pattern collected in 2013, as this structure correctly matched the predicted crystal structure. The refined structure is in very good agreement with the predicted structure, demonstrating the excellent results of crystal structure prediction by Vasileiadis. This study demonstrates the value of combining computational and experimental methods, especially in screening for polymorphs PO13 melts at 100.0°C. PO13 melts at 100.0°C. Due to the amorphous content, minor form impurities in the starting material, and overlapping thermal events, the exact heat of melting could not be determined, but the total enthalpy was about 25.6 kJ mol^{-1} and compared well with other known forms. Because of the imprecision of the heat of fusion during melting, the relative stability of PO13 relative to other forms cannot be predicted solely from the thermal data. However, samples were stored at room temperature for approximately 5 years. During this time, the relative percentage of PO13 decreased from 99% to 45% in favor of Form Y. Thus, at room temperature, Form Y is more stable than PO13.

Zhang et al. [10] discussed the metastable polymorph of the amide compound tuberculosis (pharmaceutical), of which three polymorphs are known. *Single crystal results were used to identify which known polymorph it is.* The polymorphs have structural differences in torsion angles and hydrogen bonds, as well as significant differences in lattice energies. Two new metastable polymorphs of the tuberculosis drug isoniazid, considered monomorphic for 60 years, were discovered using melt crystallization and nanoscale confinement. *The two new forms were easily distinguished from the known forms by optical microscopy, Raman spectroscopy, and powder X-ray diffraction.* For one of the new polymorphs, a single crystal structure was obtained. Melt crystallization yielded three crystalline polymorphs that were easily distinguished by polarized light microscopy. One of them was identified as Form I by Raman spectroscopy and powder X-ray diffraction. The other two were new polymorphs (forms II and III). These polymorphs were strongly temperature-dependent. The sublimation method produced needle crystals of morphology II and flake crystals of morphology I. Morphology II was identified by single-crystal X-ray diffraction; for type II, space groups were identified by single-crystal X-ray diffraction. *The calculated powder X-ray diffraction patterns are consistent with the experimental geometry patterns*; the hydrogen bonding modes for types I and II are different; the lattice energies for types I and II were *calculated using the Vienna ab initio simulation package (VASP 5.3.5) and density functional theory (DFT).* The energy cutoff for the plane wave expansion is 500 eV and was described using the Projector Augmented Wave (PAW) method. The Generalized Gradient Approximation (GGA) of the revised Perdew-Burke-Ernzerhof (rPBE) function was employed to account for electron exchange and correlation. Grimme's D2 method was used to estimate van der Waals forces. Thus, crystal morphology, growth characteristics, *PXRD, Raman spectroscopy, and single crystal X-ray diffraction* clearly indicate the presence of two new polymorphs of INZ obtained by crystallization from melt. These results indicate that melt crystallization and crystallization under nanoconfinement are promising experimental methods for screening polymorphs, which is important for the development and efficacy of active pharmaceutical ingredients. This is especially important for high-Z structures that cannot be tested by computational crystal structure prediction methods. Furthermore, these results further support McCrone's conjecture that "in general, the number of forms known for a compound is proportional to the time and money spent on the study of that compound."

Bhardwaj et al. [11] discuss more than 10 crystal polymorphs and various crystal structures with and without "crystallization solvents." There are differences in conformation and hydrogen bonding patterns among the polymorphs. Predictive calculations of crystal structures have been very difficult due to the effects of pressure and crystallization solvents. *Rietveld analysis of powder X-ray diffraction was performed using single crystal results. Calculated results were organized by creating a (lattice) energy landscape and a search was made for energetic global minima.* Polymorphs for which single crystal X-ray diffraction was not possible were submitted to powder X-ray diffraction, and similar ones were clearly distinguished by *solid-state NMR*. In the first CSP study, the $Z' = 2$ polymorph was examined for possible crystal packing in various space groups determined by the three torsion angles, energy landscapes were generated, and plausible structures were found for several polymorphs. In the next CSP study, GRACE was used for the $Z' = 2$ polymorph, sampling crystal filling in the $Z' = 1$ or $Z' = 2$ space group, which accounts for 99.9% of the distribution observed in the Cambridge Structural Database. Energy landscapes were calculated using Perdew-Burke-Ernzerhof (PBE) functions and empirical one-atom dispersion models (NP). Crystal structures were determined by Rietveld refinement of structural models using single crystal X-ray diffraction or powder X-ray diffraction combined with solid-state NMR. The relative positions of hydrogen bond donors and acceptors varied widely between polymorphs; *DSC measurements were complemented by calculations of relative lattice energies using various periodic dispersion corrected DFT methods.*

Jeziorna et al. [12] discuss the polymorphism and "co-crystallization" of meloxicam. *They use energy landscapes for crystal structure prediction calculations, characterized by DFT calculations that search for conformations. Rietveld analysis of powder X-ray diffraction is based on limited single-crystal analysis and is characterized by large differences in hydrogen bonding modes between polymorphs. Thermal analysis related to phase transitions and solid-state NMR (observation of hydrogen bonding sites) experiments, described below, were also performed.* All structures within the first 15 kJ/mol of the total energy of the global minimum (GM) structure in this energy landscape were further evaluated. Conformational searches for MLX neutral and two amphoteric ionic types are performed using Spartan 18's grid search. All rotatable bonds are included. All unique stereostructures were further subjected to DFT calculations using Gaussian 16 software, evaluated at the level of theory in B3LYP-GD3BJ/6-311G (d, p), and later evaluated in the basis set using an additional diffusion function (6-311G++ (d, p)), specifically for amphoteric The ranking of the conformational energies was confirmed to be correct with respect to the ionic form. Finally, an additional relaxation scan calculation was performed in Gaussian 16 with a small step size of 8° for this torsion in order to evaluate the energetic impact of the thiazole ring orientation. CSP calculations were performed independently for each of the selected related stereostructures; the $Z' = 1$ and $Z' = 2$ crystal structure generation steps were performed using Global Lattice Energy Explorer. Based on quasi-random searches, between 2000 and 10,000 geometrically minimized crystal structures were generated in the 10 most common crystallographic space groups. The generated structures were geometrically minimized with respect to intermolecular interactions using FIT repulsive dispersion potentials. Atom-centered multipoles up to rank 4 (generated from Gaussian calculated electron density in GDMA 2.2.11 software) and a cutoff value of 25 Å for van der Waals interactions were used for the minimization. The force-field optimized structures were evaluated for overlap based on PXRD patterns and similarity of density and energy data, with cutoff values of 0.02 g/cm^3 and 0.1 kJ/mol, respectively. The unique structures generated and optimized at this stage constitute the energy landscape of the force field. In this energy landscape, all structures within the first 15 kJ/mol total energy for global minimum (GM) structures were further evaluated. Selected MLX-predicted crystal structures were further shape-optimized with a k-point spacing of 0.07 $Å^{-1}$ and an energy cut of 600 eV using PBE-D2 level theory. In the first run of shape optimization, all atom positions were relaxed with fixed cell parameters, and in the second run, the cell parameters were also relaxed. This was done to speed up the convergence to the local minimum: a fully geometrically optimized structure in the PBE-D2 level constitutes an intermediate-level crystalline energy landscape called the DFT-D2 level. From this DFT-D2 landscape, the lowest energy structures found within the first 15 kJ/mol relative energy range of the DFT-D2 GM structure were further targeted for PBE-MBD geometry optimization.

To calculate the ^{13}C shielding constants MLX-I, MLX-IV, and SDMSO, *they used the reconstructed experimental crystal structures as a starting point and optimized the geometry with CASTEP using the same approach as the structures generated with the CSP NMR shielding constants to support ^{13}C resonance assignments. For the optimized structure, the structure was calculated with the GIPAW approach*; based on the analysis of the CSP energy landscape, new crystallization experiments were devised and performed to obtain polymorphic crystals; the lattice and intramolecular energies of crystals containing neutral and charged species of PRX, a molecule closely related to MLX The implications of the comparison of the lattice and intramolecular energies of crystals containing neutral and charged species of PRX, a molecule closely related to MLX, are also briefly discussed. Usually, when considering crystallization from a solvent, evaluating the energies of chemical species present in solution can provide some clues as to possible conformations in the crystal. In fact, zwitterionic species tend to have higher gas-phase energies than neutral species, justifying the omission of such studies. As a result, the crystal lattice energy of the charged species is 17% higher than that of the neutral species, and the crystal structure of the zwitterionic species has an intramolecular energy of more than 40 kJ/mol higher than that of the neutral molecules. This is important not only for solution crystallization studies but also for CSP calculations, which must take into account that the intramolecular energy of the zwitterionic species is higher than that of the neutral case.

Some examples of studies feature methods to obtain polymorphic crystals: Askin et al. [13] *obtained a new polymorph* of olanzapine, a pharmaceutical compound with *more than 60 reported crystal structures including polymorphs and crystal water*, by crystallizing it in a polymeric PVP solution. The dimer structure in the crystal is characteristic. Melting and crystallization were directly observed by simultaneous thermal analysis and powder XRD measurements. The results of lattice energy calculations are discussed as landscape. A new polymorph of olanzapine, type IV, was identified and isolated; CSP accurately predicted the presence of type IV, which had not been obtained experimentally prior to this study because solution-based crystallization methods form olanzapine dimers and further polymorphs based on the SC 0 motif. Crystallization from the amorphous phase of solids allows crystallization of different structures, which is thought to be due to the prevention of dimer formation. *DSC-PXRD* has been important in confirming the presence of new morphologies. On the other hand, Braun et al. [14] reported surface-induced polymorphism of the antiepileptic drug phenytoin. In surface adsorption, it is hypothesized that polymorphism can occur due to a decrease in local entropy at the surface due to the interaction between the adsorbed molecule and the substrate. The polymorphs were screened from the landscape of lattice energy calculations (information that polymorphs could be obtained). Two crystalline polymorphs were obtained when crystallized on the surface of silicon wafers. *These were confirmed by grazing incidence X-ray diffraction (scattering vector values) corresponding to the surface, rather than by the usual powder XRD measurements of bulk samples.* Detailed features and differences in molecular conformation and crystal packing (hydrogen bonding mode differences) are discussed. The structure of phenytoin in thin films is difficult to assess for many reasons; a combination of CSP calculations, GIXD experiments, and Raman spectroscopy allowed us to elucidate the structure of the thin film form (Form II) of phenytoin. The results suggest that differences in the steric H-bonding network are likely the reason for the faster dissolution of the thin film form (Form II) induced at the surface, as opposed to the bulk form, Form I. In general, CSP-based structural solution methods can provide information on the packing arrangement of organic molecules with limited experimental data. This can be applied to low-quality data such as GIXD data for thin films, X-ray diffraction for single crystals, neutron diffraction, electron diffraction, etc., minimizing the requisite for high-quality crystal growth. Regarding entropy in normal crystal polymorphs, there is also a study of benzamide by Fellah et al. [15]. A new highly metastable benzamide Form IV polymorph was found by melt crystallization in parallel with growth under nanoconfinement in very small pores. Under nanoscale configurations, the melting behavior of the more stable benzamide Form III crystallized in larger pores was strongly affected by pore size. *The interfacial energies calculated by the Gibbs-Thomson equation were consistent with those obtained by the semiempirical model.* A number of candidate structures for benzamide IV were obtained by a combination of powder X-ray diffraction analysis and computational methods. Several possible disorder models were narrowed down based on a two-dimensional parquet-like tiled tape consisting of four benzamide molecules. The disorder can be explained by various stacking defect configurations corresponding to random displacement and orientation of the tape. Taken together, these observations reveal that although the crystallographic complexity of such a simple molecule is impressive, new methods for determining solid-state structure are within reach. Moreover, this analysis suggests that many structures that might have been abandoned due to overarching defects such as low crystallinity, twinning, and disorder can be resolved. Yao et al. [16] discovered two new polymorphs of piroxicam (forms VI and VII) by melt crystallization and crystal structure prediction (CSP). They grew high-quality single crystals from molten microdroplets and analyzed their structures by X-ray diffraction. Melt crystallization is an important complement to solution analysis in polymorph discovery because it is thermodynamically less stable than the known polymorphs, which are mainly obtained by solution analysis. One thing has been confirmed. Piroxicam was reported to be thermally unstable and unsuitable for melt crystallization, but this problem was circumvented by exposing the molten droplets to high temperature for a short time.

If there are differences in hydrogen bonding modes between crystal polymorphs, a method that can *combine spectroscopic experiments with XRD would be promising to detect them. Vibrational spectroscopy is available for hydrogen bonding, but solid-state NMR is more useful when compared to computational chemistry,* and Mathew et al. [17] used a combined experimental and computational approach to distinguish between different polymorphs of the drug aspirin. In this method, DFT calculations of NMR parameters and solid-state NMR experiments using isotopes of natural abundance are combined with a DFT-based crystal structure prediction method, the ab initio random structure search method (AIRSS method), to predict polymorphic structures. Prediction with high accuracy AIRSS was used to predict the crystal structures of two polymorphs of aspirin, and information from powder XRD was also used. The root-mean-square deviation of the experimental and calculated proton chemical shifts was used to identify the polymorphs present in the experimental samples. This was successful despite the very similar molecular environments in the crystals of the two polymorphs; Smalley et al. [18] investigated the crystal structure of a new polymorph (β-polymorph) of the typical amino acid L-tyrosine by a combined analysis of three-dimensional electron diffraction data and powder X-ray diffraction. Three-dimensional electron diffraction data have been shown to be a useful structure determination method for reliable unit cell determination and space group assignment. The use of solid-state NMR and AIRSS resulted in an approach that greatly outperformed attempts at powder XRD analysis, which could not even be indexed (due to weaker XRD peak intensities) due to randomly oriented samples. A new polymorph of L-tyrosine (β polymorph) was prepared by crystallization from the gas phase, and the crystal structure was determined by combined analysis of 3D-ED and PXRD data and cyclic DFT-D calculations. The final structure of the β polymorph obtained from Rietveld purification was confirmed by (i) cyclic DFT-D geometry optimization, which confirmed that the crystal structure was very close to the minimum on the energy landscape; (ii) isotropic ^{13}C NMR for the crystal structure calculated using DFT-D methods chemical shifts and excellent agreement with values observed in experimental high-resolution solid-state ^{13}C NMR spectra. Furthermore, DFT-D calculations at the PBE0-MBD level suggest that the crystal structure of the β polymorph determined here is metastable compared to the previously reported crystal structure of the α polymorph. Structure prediction calculations using the AIRSS method generated five promising polymorphic crystal structures for L-tyrosine. The lowest energy structure was the alpha polymorph, the beta polymorph was at about 4 kJ mol^{-1} higher energy, and the other three predicted structures were at more than 11 kJ mol^{-1} higher energy than the alpha polymorph. In determining the crystal structure of materials that are difficult to characterize structurally by single-crystal XRD, the important advantages of using a combination of these methods within a robust protocol that incorporates consideration of periodic DFT-D calculations and solid-state NMR data, in addition to the combination of 3D-ED and PXRD data are highlighted. This has expanded our understanding of the structural properties of L-tyrosine in its crystalline state.

There is also a paper by Pawlak et al. [19] on *combining solid-state NMR with experiments and crystal structure prediction calculations for single crystals, which is more reliable than powder XRD.* The complementary application of CSP and solid-state NMR has significant advantages. Furthermore, while CSP describes well the range of thermodynamically preferred crystal structures, even the most robust algorithms do not take into account the static/dynamic molecular disorder that is often observed in experimental techniques. Solid-state NMR can be easily applied to systems that exhibit conformational flexibility and different molecular interactions. Thus, they demonstrate the synergistic effects of experimental solid-state NMR and single-crystal X-ray diffraction, as well as thermal analysis and crystal structure prediction techniques based on DFT calculations, in the analysis of systems that form polymorphs, undergo phase transitions, and local dynamic processes. One of the simplest and most common applications of *solid-state NMR is the use of one-dimensional ^{13}C CP/MAS experiments to determine the number of non-equivalent molecules, or crystallographic Z′ values.* This is the first step in a strategy called NMR crystallography, the core of which is the GIPAW method. The GIPAW method is a breakthrough development in the theoretical prediction of NMR parameters for solid materials, and this approach can be readily applied to provide experimental validation of CSP solutions. Specifically, we show whether the possible solutions identified by the CSP strategy are consistent with the experimental form. In this study, we present for the first time the X-ray diffraction structure and MAS NMR characterization of a new room temperature polymorph of TFM (labeled TFM RT) recently reported by Gunnam et al. and a low-temperature TFM polymorph (TFM LT) at −40°C that undergoes thermal transition to the TFM RT polymorph. The phase transition is clearly visible in the DSC plots and is reversible. Crystals mounted on a goniometer head can be cooled and heated without cracking, producing diffraction patterns characteristic of TFM LT and TFM RT lattices, respectively. This reversible process occurs by varying the Z′ value while maintaining the crystal system, and the two forms have different lattice parameters and exhibit different reflection patterns. Reflections per second corresponding to TFM LT lattice direction a, both structures vanishing in TFM RT diffraction images, for the same type of crystal Reflections per second corresponding to TFM LT lattice direction a, both structures vanishing in TFM RT diffraction images, were determined based on diffraction experiments performed on the same type of crystal, highlighting that The differences between TFM LT and TFM RT will be discussed, focusing on the change in aromatic ring orientation related to the C4-NH-C5-C6 torsion angle. In the room temperature bimolecular structure, up to 50% disorder is modeled between torsion angles of −7.8° and 28.6°. These observations are consistent with the broad energy minima observed in the DFT calculations when the C4-NH-C5-C6 torsion angle is varied; PISEMA solid-state NMR experiments show that the static limits of the aromatic CH portion are 75% and 51% at 20 and 40°C, respectively, while the C-H The difference in crystal lattice energies between TFM LT and TFM RT is up to 2.4 kJ/mol, as calculated using DFT-D. This small difference supports the reversible conversion between both forms observed experimentally. The TFM RT polymorph also exhibits molecular disorder of the CF 3 group, and thus, the disorder of the two parts of the molecule may be correlated. The conversion from TFM LT and TFM RT appears to be driven by the same entropy preference as previous studies. The transformations from TFM LT and TFM RT appear to be driven by the same entropy preference as in previous studies. It is interesting to compare the TFM cases here: impairment in TFM RT is associated with a change in Z′ value from 2 to 1 compared to TFM LT in the case of Szell et al. However, there were clear differences in the calculated PXRD patterns that could not be explained by thermal expansion alone, and the ^{13}C chemical shifts observed by solid-state NMR were different at all sites, not just those showing disorder. Furthermore, the crystal structure prediction procedure, which does not account for disorder in the output format, explained both the experimental TFM polymorphism in the global minimum region of energy. In conclusion, this study demonstrates that DFT calculations and CSP, in combination with experiments such as DSC, X-ray diffraction, and MAS NMR, can be used to investigate and understand the role of solid-state states, especially dynamics, in pharmaceutical molecules.

3. Three-dimensional electron diffraction for crystal structure analysis

Methods for powder structure analysis of organic compounds were reviewed, including additional techniques (other measurements and computational chemistry) and prediction of (polymorphic) crystal structures. However, the biggest experimental difficulty is that the (single) crystal is small, and there are great expectations that advances in equipment such as electron diffraction will be able to overcome this issue. Recent advances in instrumentation and technology [20, 21] have made it possible to use electron diffraction for crystallography (3-D structural analysis) rather than traditional electron microscopy (which has been used for morphological observation of metal compound nanocrystals.). In 1994, Vincent and Midgley developed a new data acquisition tool, *precession electron diffraction (PED)*, which alleviates some of these problems. In 2007, Kolb et al. used a TEM goniometer to tilt a substrate in a series of *discrete angular steps (ADT)* and proposed to collect a series of tomographic diffraction patterns. ADT is intentionally designed so that most diffraction patterns are collected in the off zone, providing ideal conditions for observing quasi-kinematic scattering; ADT was rapidly combined with beam precession by Mugnaioli et al. Although PEDT required special external hardware to achieve beam precession, structural PEDT was the first electron diffraction technique to gain some traction as a general method for structure elucidation. Hovmöller, Zou, et al. also devised a method to slice the reciprocal lattice space into finer slices by supplementing the coarse mechanical tilting with electron beam tilting. This method, called *Rotational Electron Diffraction (RED)*, utilizes custom software to collect data at very fine angular steps ($\Delta\eta < 0.1°$), surpassing the accuracy of TEM goniometers. These developments paved the way for continuous rotation, arguably the most impactful methodological advance in *3D electron crystallography. Continuous rotation* was formulated almost in parallel by Nederlof et al. in 2013 and Nannenga et al. in 2014, respectively; unlike PEDT and RED, no auxiliary hardware or software is strictly necessary to perform continuous rotation ED, and most commercially available TEMs can collect continuous rotation data with little or no reconstruction. In this technique, the irradiated crystal is periodically sampled in reciprocal space at periodic intervals while being rotated in one direction around the axis of the TEM goniometer.

A review paper by Simoncic et al. [22] was written for crystallographers unfamiliar with 3D electron diffraction (i.e., XRD users). There are also solutions to some of the problems and challenges in the analysis of crystal structures from salen-type nickel(II) complexes and powder XRD data. *Neutrons* are scattered by their interaction with nuclei, whereas electron beams, like X-rays, are scattered by their interaction with the electron clouds of atoms and nuclei (electrons are electronegative, so they interact electrostatically with the positive site). However, electron diffraction, which can measure single crystals of even smaller size (thinner than 1 μm thick) than X-ray diffraction, is useful when single crystal growth is difficult. Since electron beams have a different wavelength range and diffraction angle than X-rays, the *problem of overlapping diffraction peaks* in samples with long lattice constants (which hinders indexing and analysis) can be avoided. Electron diffraction requires only a small amount of sample and is not easily affected by small amounts of *impurities (multiple crystalline phases mixed together)*.

From the work of Zhou et al. [23], we illustrate the advantages of electron diffraction, keeping in mind other problems of single crystal size in salen-type nickel(II) complexes (low crystallinity and presumably also anisotropic molecular orientation ordering). Some organic compounds, such as covalent organic frameworks (COFs), have low crystallinity and pores in the crystal. Of course, the resolution of diffraction is low. Such macromolecular crystalline compounds are affected by a *mixture of kinetically and thermodynamically favorable products*, the difficulty of obtaining large single crystals, and the influence of *structural disorder on the Bragg reflection*. Crystal structure analysis is difficult because of the susceptibility to radiation damage. This is a successful example of structural analysis by 3D electron diffraction, combining high-quality single-crystal data obtained separately with refinement calculations using simulated annealing (rather than conventional methods). The use of *SA (simulated annealing)* enables us to obtain a structural model that is comparable to the reference model, even though it contains skeletal distortions. This indicates the possibility of solving low crystallinity COF structures with SA. On the other hand, structural analysis using low-resolution data is difficult with the dual-space method, the charge inversion method, and the classical direct method. In addition, we performed simulations with different resolution data and investigated how they affect the structural elucidation by SA. The results show that a resolution of 1.5 Å or higher produces a good structural model. However, when the resolution is reduced below 2.0 Å, distortion increases significantly and prevents meaningful interpretation of the framework structure. Since crystallization is not a problem limited to COFs, it is hoped that the use of SA for low-resolution 3DED data will allow the determination of the structures of other framework materials, such as metal-organic frameworks (MOFs), hydrogen bond frameworks (HOFs), and zeolites.

4. Conclusions

In a metal complex containing a specific nickel(II) metal ion, the single crystal becomes brittle when it exhibits a structural phase transition. However, structural analysis from powder XRD data often does not go smoothly. Even with single-crystal size and crystallinity issues, the potential for beams to reach real-space goals as well as reciprocal space data (diffraction patterns) would improve things. It is also expected that the real crystal structure of the nickel(II) complex mentioned above will eventually be solved. If it is allowed to "fit" the prediction results to the XRD pattern, the current style of "solving" the crystal structure (overcoming the so-called phase problem and obtaining the atomic arrangement in the real space from the measured information in the reciprocal space) may be in danger.

Acknowledgements

The authors thank Dr. Dohyun Moon (Pohang Light Source II, Korea) for the measurement of XRD and Dr. Akane Samizo (Rigaku Cooperation, Japan) for electron diffraction. We apologize for not being able to analyze the crystal structures.

Conflict of interest

The authors declare no conflict of interest.

Author details

Yuya Higashi, Rin Tsuchiya, Diasuke Nakane and Takashiro Akitsu*
Faculty of Science, Department of Chemistry, Tokyo University of Science, Tokyo, Japan

*Address all correspondence to: akitsu2@rs.tus.ac.jp

References

[1] Ishikawa T, Honda A, Miyamura K. Cold crystallization in mixed systems of alkyl-derivatized Schiff-base nickel(II) complexes. Bulletin of the Chemical Society of Japan. 2022;**95**:1747-1752. DOI: 10.1246/bcsj.20220276

[2] Ishikawa T, Honda A, Miyamura K. Effects of alkyl chain length on the cold crystallization of Schiff-base nickel(II) complexes. CrystEngComm. 2022;**24**:5900-5906. DOI: 10.1039/D2CE00305H

[3] Siegler MA, Lutz M. Ni(salen): A system that forms many solvates with interacting Ni atoms. Crystal Growth & Design. 2009;**9**:1194-1200. DOI: 10.1021/cg801109n

[4] Adhikari R, Kotbagi TV, Shaughnessy KH, Shrestha AB, Sherwood J, Bao Y, et al. How addition of a nickel cyclohexyl-salen complex impacts a one-pot synthesis of nickel/hierarchically porous carbon monolith catalyst. Journal of Sol-Gel Science and Technology. 2021;**97**:106-116. DOI: 10.1007/s10971-020-05436-3

[5] CCDC: SAENNI.Shkolnikova LM, Yutal EM, Shugam EA, Voblikova VA. 1970

[6] Tang C. {2,2′-[Cyclohexane-1,2-diylbis(nitrilomethylidyne)]diphenolato}nickel(II). Acta Cryst. 2009;**E65**:m317. DOI: 10.1107/S1600536809006084

[7] Hoshina G, Tsuchimoto M, Ohba S. [N,N′-Bis(salicylidene)-1,2-diphenyl-(RS,SR)-1,2-ethanediaminato]nickel(II). Acta Cryst. 2000;**C56**:e122. DOI: 10.1107/S0108270100003255

[8] Watts TA, Price LS, Price SL, Niederberger SM, Bertke JA, Swift JA. The crystal structure of 5-aminouracil and the ambiguity of alternative polymorphs. Israel Journal of Chemistry. 2021;**61**:590-597. DOI: 10.1002/ijch.202100062

[9] Gushurst KS, Nyman J, Boerrigter SXM. The PO13 crystal structure of ROY. CrystEngComm. 2019;**21**:1363-1368. DOI: 10.1039/C8CE01930D

[10] Zhang K, Fellah N, Shtukenberg AG, Fu X, Hu C, Ward MD. Discovery of new polymorphs of the tuberculosis drug isoniazid. CrystEngComm. 2020;**22**:2705-2708. DOI: 10.1039/D0CE00440E

[11] Bhardwaj RM, McMahon JA, Nyman J, Price LS, Konar S, Oswald IDH, et al. A prolific solvate former, galunisertib, under the pressure of crystal structure prediction, produces ten diverse polymorphs. American Chemical Society. 2019;**141**(35):13887-13897. DOI: 10.1021/jacs.9b06634

[12] Jeziorna A, Paluch P, Zając J, Dolot R, Dudek MK. Crystallization of elusive polymorphs of meloxicam informed by crystal structure prediction. Crystal Growth & Design. 2023;**23**(8):5998-6010. DOI: 10.1021/acs.cgd.3c00559

[13] Askin S, Cockcroft JK, Price LS, Gonçalves AD, Zhao M, Tocher DA, et al. Olanzapine form IV: Discovery of a new polymorphic form enabled by computed crystal energy landscapes. Crystal Growth & Design. 2019;**19**(5):2751-2757. DOI: 10.1021/acs.cgd.8b01881

[14] Braun DE, Rivalta A, Giunchi A, Bedoya-Martinez N, Schrode B, Venuti E, et al. Surface induced phenytoin polymorph. 1. Full structure

solution by combining grazing incidence X-ray diffraction and crystal structure prediction. Crystal Growth & Design. 2019;**19**(11):6058-6066. DOI: 10.1021/acs.cgd.9b00857

[15] Fellah N, Shtukenberg AG, Chan EJ, Vogt-Maranto L, Xu W, Li C, et al. Disorderly conduct of benzamide IV: Crystallographic and computational analysis of high entropy polymorphs of small molecules. Crystal Growth & Design. 2020;**20**(4):2670-2682. DOI: 10.1021/acs.cgd.0c00096

[16] Yao C, Guzei IA, Jin Y, Ruan S, Sun G, Gui Y, et al. Polymorphism of piroxicam: New polymorphs by melt crystallization and crystal structure prediction. Crystal Growth & Design. 2020;**20**(12):7874-7881. DOI: 10.1021/acs.cgd.0c01165

[17] Mathew R, Uchman KA, Gkoura L, Pickard CJ, Baias M. Identifying aspirin polymorphs from combined DFT-based crystal structure prediction and solid-state NMR. Magnetic Resonance in Chemistry. 2020;**58**:1018-1025. DOI: 10.1002/mrc.4987

[18] Smally CJH, Hoskyns HE, Hughes CE, Johnstone DN, Willhammar T, Young MT, et al. A structure determination protocol based on combined analysis of 3D-ED data, powder XRD data, solid-state NMR data and DFT-D calculations reveals the structure of a new polymorph of L-tyrosine. Chemical Science. 2022;**13**:5277-5288. DOI: 10.1039/D1SC06467C

[19] Pawlak T, Sudgen I, Bujacz G, Iuga D, Brown SP, Potrzebowski MJ. Synergy of solid-state NMR, single-crystal X-ray diffraction, and crystal structure prediction methods: A Case Study of Teriflunomide (TFM). Crystal Growth & Design. 2021;**21**(6):3328-3343. DOI: 10.1021/acs.cgd.1c00123

[20] Gruene T, Mugnaioli E. 3D electron diffraction for chemical analysis: Instrumentation developments and innovative applications. Chemical Reviews. 2021;**121**:11823-11834. DOI: 10.1021/acs.chemrev.1c00207

[21] Saha A, Nia SS, Rodriguez JA. Electron diffraction of 3D molecular crystals. Chemical Reviews. 2022;**122**:13883-13914. DOI: 10.1021/acs.chemrev.1c00879

[22] Simonics P, Romeijin E, Hovestreydt E, Steinfeld G, Santiso-Quinones G, Merkelbach J. Electron crystallography and dedicated electron-diffraction instrumentation. Acta Cryst. 2023;**E79**:410-422. DOI: 10.1107/S2056989023003109

[23] Zhou G, Yang T, Huang Z. Structure determination of a low-crystallinity covalent organic framework by three-dimensional electron diffraction. Communications Chemistry. 2023;**6**:116. DOI: 10.1038/s42004-023-00915-4

Chapter 7

Characterization of Schiff Base Ligand and Its Metal Complexes

Mohammed Umar Adaji, Moses Saviour Iorungwa and Olalekan Wasiu Salawu

Abstract

Schiff bases, derivatives born from the condensation of primary amines with carbonyl compounds, have long captured the attention of chemists due to their intriguing structural, biological, and coordination properties. The versatility of these compounds is evident in their wide array of applications ranging from medicinal chemistry to coordination-driven processes and material science. This chapter provides an exhaustive exploration into the synthesis, characterization, and applications of Schiff base ligands and their metal complexes. Emphasis is laid on the myriad ways Schiff bases can be tailored to design metal complexes with specific properties, opening doors to targeted applications. Through a blend of fundamental principles, modern characterization techniques, and real-world applications, readers will gain a holistic understanding of Schiff base chemistry, its interplay with metals, and its bearing on various scientific and industrial domains. This comprehensive guide serves as a testament to the adaptability and significance of Schiff bases in contemporary chemical research and development.

Keywords: Schiff base, azomethine, coordination, complexation, metal complexes, and characterization

1. Introduction

Schiff bases, named after Hugo Schiff who first reported them in the nineteenth century, are a class of organic compounds characterized by a functional group containing a carbon-nitrogen double bond with the nitrogen atom connected to an aryl or alkyl group but not hydrogen. Their general structure is RHC=N-R′, where R and R′ can be a variety of substituents. These compounds are primarily synthesized by the condensation of primary amines with carbonyl compounds, typically aldehydes or ketones. The importance of Schiff bases transcends their straightforward synthesis. They have been a subject of scientific interest due to their diverse range of biological activities, including anticancer, antiviral, antibacterial, antifungal, anti-inflammatory properties, and nematicidal activities [1]. This has led to their study in the realm of medicinal chemistry and their application in drug design.

In the field of coordination chemistry, Schiff bases have carved a niche for themselves as versatile ligands. They can chelate to metal ions through the nitrogen atom of the azomethine group, forming stable ring structures. The resulting metal complexes often exhibit remarkable properties, which make them suitable for a wide array of applications. They can act as catalysts, sensors, and even as materials with magnetic or photophysical properties.

The study of Schiff base metal complexes is not just about understanding their structure, but also about tapping into their potential applications for clarity. Whether it is designing a new catalyst for a challenging chemical reaction or developing a metal-based drug with fewer side effects, Schiff base ligands, and their metal complexes offer a rich tapestry of possibilities.

Moreover, the ease of structural modification in Schiff bases, due to their synthetic accessibility, allows for tuning of the electronic and steric properties of the ligand. This paves the way for designing metal complexes with tailor-made properties, catering to specific applications.

In this chapter, we will journey through the realm of Schiff base ligands from their synthesis to their complexation with metals, diving deep into their characterization, and understanding the nuances that make them special in the world of coordination chemistry.

2. Synthesis of Schiff bases

Schiff bases are condensation products of primary amines and carbonyl compounds, and they were discovered by a German chemist, Nobel Prize winner, Hugo Schiff [2]. Structurally, Schiff base (also known as imine or azomethine) is an analog of a ketone or aldehyde in which the carbonyl group (C=O) has been replaced by an imine or azomethine group (**Figure 1**) [3]. A Schiff base is a type of chemical compounds containing a carbon-nitrogen double bond as functional group, where the nitrogen atom is connected to aryl group or alkyl group (R) but not hydrogen. The Schiff base is synonymous with an azomethine. These compounds were named after Hugo Schiff on honor and have the following general structure:

Where R stands for a phenyl or alkyl group which makes the Schiff base a stable imine. This kind of ligand is able to coordinate metal ions through the imine nitrogen and another group, usually linked to the aldehyde. The chemists still prepare Schiff bases and nowadays active and well-designed Schiff base ligands are considered "privileged ligands" [4]. The bridged Schiff bases have the following structure, which contains many functional groups able to change according to the purpose required (**Figure 2**).

where R′ = H or alkyl group, R″ = phenyl or substituted phenyl, R‴ = alkyl or aryl group.

In fact, Schiff bases are able to stabilize many different metals in various oxidation states, controlling the performance of metals in a large variety of useful catalytic

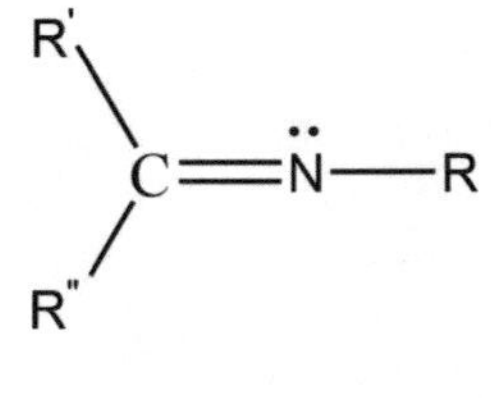

Figure 1.
Schiff base.

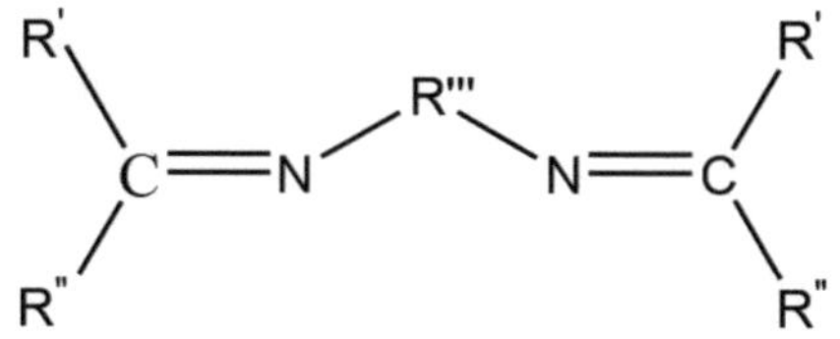

Figure 2.
Bridged Schiff base.

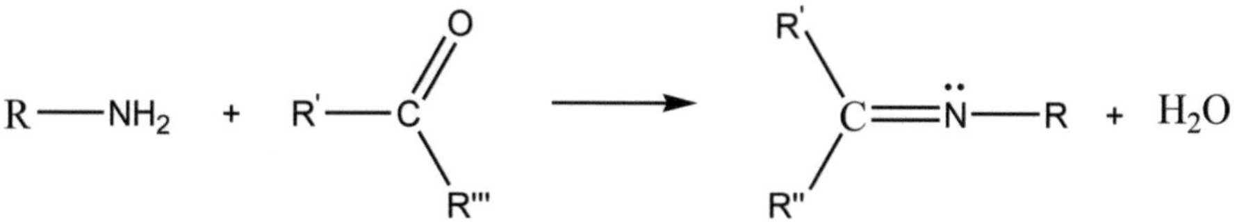

Figure 3.
Formation of Schiff base by condensation reaction.

transformations. Most commonly Schiff bases have NO or ONO donor groups, but the oxygen atoms can be replaced by sulfur or nitrogen atoms.

It is usually formed by condensation of an aldehyde or ketone with a primary amine according to the following **Figure 3**.

Where R may be an alkyl or an aryl group. Schiff bases that contain aryl substituent are substantially more stable and more readily synthesized, while those, which contain alkyl substituent are relatively unstable. Schiff bases of aliphatic aldehydes are relatively unstable and readily polymerizable [5]. While those of aromatic aldehydes having effective conjugation are more stable [2].

Schiff bases, commonly represented by the general formula RHC=N-R′, arise from the condensation of a primary amine with a carbonyl compound, typically an aldehyde or ketone. The synthetic process is relatively straightforward and often proceeds under mild conditions, making it a staple in organic synthesis laboratories.

2.1 General Synthetic Procedure

Materials:

- A primary amine: RNH_2
- An aldehyde or ketone: R′CHO or R′CO

Procedure:

1. To a solution of the primary amine in a suitable solvent (e.g., ethanol), add the aldehyde or ketone.
2. The mixture is then stirred at room temperature or by heating.
3. After a few hours (depending on the specific reactants), the reaction typically reaches completion. This can be monitored using thin-layer chromatography (TLC) or other suitable analytical techniques.

4. The Schiff base can then be isolated by filtration, if it is solid, or through solvent removal if it remains in the solution. It is then typically purified using recrystallization.

2.2 Mechanism of formation

The mechanism for the formation of Schiff bases is a classic example of nucleophilic addition followed by a dehydration step:

1. *Nucleophilic Addition*: The primary amine attacks the carbonyl carbon of the aldehyde or ketone. This results in the formation of a carbinolamine intermediate.

2. *Dehydration*: The carbinolamine intermediate loses a molecule of water to form the desired Schiff base, characterized by its C=N azomethine bond.

2.3 Factors affecting Schiff base formation

Several factors can influence the formation of Schiff bases:

1. *Catalysis*: The addition of acid or base can catalyze the reaction. Acid catalysis typically enhances the electrophilicity of the carbonyl carbon, while base catalysis can increase the nucleophilicity of the amine.

2. *Solvent*: Polar solvents, particularly protic ones, tend to favor the reaction. Common solvents include alcohols and water, although the choice can depend on the solubility of the starting materials [6].

3. *Substituents on the Reactants*: Electron-withdrawing groups on the carbonyl compound can make it more susceptible to nucleophilic attack, thereby speeding up the reaction.

2.4 Variability and versatility

The ease of synthesis and the plethora of available primary amines and carbonyl compounds mean that a vast array of Schiff bases can be synthesized, each with its unique properties. By changing the R and R′ groups, chemists can tailor-make Schiff bases for specific applications, whether it is to bind a particular metal ion or to exhibit a certain biological activity.

In summary, Schiff bases are a cornerstone in the realm of organic and coordination chemistry, serving as a bridge between the two worlds. Their synthesis is emblematic of the elegant simplicity that can sometimes underpin the most profound chemical phenomena.

3. Characterization method

3.1 Spectroscopic methods

Schiff bases and their metal complexes often exhibit unique spectroscopic features that allow researchers to gain insights into their structure, bonding, and other characteristics. The following are some of the most pertinent spectroscopic techniques used in the study of Schiff base ligands and their metal complexes:

3.1.1 UV-Visible spectroscopy

- *Principle*: Studies the absorption of ultraviolet (UV) and visible light, which results in electronic transitions in molecules.

- *Applications*: Used to determine conjugation in organic molecules, assess the electronic environment in metal complexes, and measure concentrations *via* Beer's Law.

- *Relevance*: Schiff bases and their metal complexes typically show distinct UV-visible absorption bands. The electronic transitions often include $\pi \rightarrow \pi^*$ transitions (due to the conjugated system) and $n \rightarrow \pi^*$ transitions (from nonbonding electron pairs). When Schiff bases coordinate to metal ions, new bands associated with metal-to-ligand charge transfer (MLCT) or d-d transitions (in transition metals) may arise. The position, intensity, and nature of these bands can provide information about the electronic environment around the metal center and the nature of the metal-ligand bond.

3.1.2 Infrared (IR) spectroscopy

- *Principle*: Infrared radiation absorption results in vibrational transitions in molecules. Each functional group has a characteristic vibration frequency.

- *Applications*: Widely used for identification of functional groups in organic compounds and studying molecular vibrations.

- *Relevance*: The formation of Schiff bases involves the creation of a C=N bond, which exhibits a characteristic stretching frequency in the IR spectrum. This C=N stretching frequency is a crucial marker for the successful formation of a Schiff base. Upon metal coordination, this frequency might shift, indicating the binding mode of the ligand to the metal ion. Other characteristic peaks related to the ligand can also be observed and monitored for changes upon metal coordination.

3.1.3 Nuclear magnetic resonance (NMR) spectroscopy

- *Principle*: Explores the magnetic properties of certain atomic nuclei (e.g., 1H, 13C) under an external magnetic field.

- *Applications*: Provides detailed information about molecular structure, dynamics, reaction state, and environment of molecules.

- *Relevance*: 1H NMR spectroscopy can give detailed insights into the protons' environment in the Schiff base ligand. Changes in chemical shifts upon metal complexation can hint at the areas of the ligand involved in binding, especially if certain protons come close to paramagnetic metal centers. 13C NMR can provide similar insights for carbons, especially in elucidating the coordination environment.

3.1.4 Electron spin resonance (ESR) spectroscopy (or electron paramagnetic resonance, EPR)

- *Principle*: ESR (or EPR, Electron Paramagnetic Resonance) measures the interactions of unpaired electron spins in a magnetic field.

  Relevance to Schiff bases: For Schiff base metal complexes containing transition metals with unpaired electrons, ESR can be an invaluable tool. It provides information about the electronic environment of the metal center, aiding in the identification of the oxidation state and the nature of the metal-ligand bond.

- *Relevance*: ESR is especially relevant for metal complexes of Schiff bases where the metal is in a paramagnetic state (having unpaired electrons). ESR can give details about the number and nature of unpaired electrons, helping in the identification of the oxidation state, the nature of the metal-ligand bond, and the electronic environment of the metal center.

3.1.5 X-ray photoelectron spectroscopy (XPS)

- *Principle*: XPS investigates the photoelectric effect where core electrons are emitted upon irradiation with X-rays. The measured kinetic energy of these electrons, combined with the known energy of the X-rays, reveals the binding energy of the electrons. Specific elements have characteristic binding energies, and the slight shifts in these energies can hint at their chemical state or environment.

- *Applications;*

1. Identification of elements: XPS can confirm the presence of elements in Schiff base ligands, such as nitrogen from the imine group or other heteroatoms.

2. Coordination chemistry: For Schiff base metal complexes, XPS can determine the oxidation state of the coordinated metal and provide insights into the nature of the metal-ligand bond.

3. Chemical environment: Variations in the binding energy can suggest different environments or states of the same element within the Schiff base. For example, nitrogen in a free Schiff base ligand might have a slightly different binding energy than nitrogen coordinated to a metal. Therefore, XPS may also indicate specific C=N's C or N as both 1 and 3.

4. Purity and surface analysis: XPS can detect surface contaminants or degradation products, ensuring the purity of synthesized Schiff base compounds or their complexes.

5. Interactions and binding modes: For Schiff bases used in surface modifications or as linkers in materials, XPS can help in understanding their binding mode and interaction with surfaces or other entities.

6. In essence, for Schiff bases and their associated metal complexes, XPS serves as a pivotal tool, offering detailed information on elemental composition, oxidation states, and chemical environments, crucial for their thorough characterization and application in various domains.

3.1.6 Mass spectrometry (MS)

- *Principle*: Measures the mass-to-charge ratio of ions. A sample is ionized, and the resulting ions are separated based on their mass-to-charge ratio.

- *Applications*: Used to determine molecular weight, elemental composition, and structural information of molecules.

- *Relevance*: MS can be used to confirm the molecular weight and structure of a synthesized Schiff base or its metal complex. Fragmentation patterns can also help elucidate the structure, especially in cases of complex ligands or metal assemblies.

In summary, the study of Schiff base ligands and their metal complexes often necessitates a multipronged spectroscopic approach. Each technique can provide specific insights, and together they offer a comprehensive picture of the system under study. When characterizing a new Schiff base or its metal complex, researchers typically employ a combination of the above techniques to validate their findings and gain a holistic understanding of the molecule's nature.

3.2 X-ray diffraction (XRD) and its principle

The technique of X-ray diffraction (XRD) is employed in the study of crystalline materials. X-rays are diffracted or scattered by the crystal lattice when they strike a crystalline sample. This diffraction pattern can be examined to ascertain the substances molecular structure because it provides details about the locations of the atoms within the crystal [7].

3.2.1 Application to Schiff bases and their metal complexes

Determination of Molecular Geometry: Using XRD, one can precisely determine the geometry surrounding a central metal atom or ion, be it tetrahedral, square-planar, octahedral, etc. This is particularly important for Schiff base metal complexes as the geometry can affect their magnetic properties, reactivity, and other properties.

Coordination Mode: Schiff bases have the ability to bind to metal centers in a variety of ways by acting as multidentate ligands. If the Schiff base binds in a chelating way, forming rings with the metal center, XRD can demonstrate whether it functions as a bidentate, tridentate, or even polydentate ligand.

Bond Lengths and Angles: Bond lengths and angles can be measured using XRD. This can reveal details about the strain present in the complex, as well as the type of metal-ligand bond [8]—whether it is more covalent or ionic—in Schiff base metal complexes. Therefore, bond length of C=N in Schiff bases is shorter than that of C-N generally.

Presence of other ligands or molecules: XRD can detect the presence of additional ligands in the metal complex, as well as any solvent molecules or counterions that may be connected to it. These may have an impact on the complex's stability and characteristics.

Intramolecular and Intermolecular Interactions: XRD can show interactions between and within complexes in the crystal lattice, including hydrogen bonding and π-π stacking. The Schiff base metal complex's solubility, stability, and other characteristics may be influenced by these interactions.

Verification and characterization: X-ray diffraction (XRD) offers unambiguous evidence of the atomic structure and arrangement, whereas other techniques, such as infrared spectroscopy or nuclear magnetic resonance (NMR), may indicate the formation of a Schiff base and its complexation with a metal.

3.3 Magnetic susceptibility and its basics

A substance's ability to become magnetized in the presence of an external magnetic field is measured by its magnetic susceptibility (χ). This characteristic provides insight into the number of unpaired electrons present in metal complexes [9].

- *Diamagnetic:* All electrons are paired. The compound is not attracted to a magnetic field. It has a magnetic susceptibility close to zero.

- *Paramagnetic: It* contains unpaired electrons. The compound is attracted to a magnetic field. The extent of attraction (or the value of χ) is directly related to the number of unpaired electrons.

Magnetic susceptibility can be used in the elucidation of Schiff base ligands and metal complexes in other to study *viz*:

1. *Metal Oxidation State and Electron Configuration:* We can determine the number of unpaired electrons in a metal and possibly its oxidation state and d-electron configuration by measuring its magnetic susceptibility. A Cu(II) ion in a similar environment (d^9 configuration) is paramagnetic with one unpaired electron, whereas a Ni(II) ion in a square planar environment (d^8 configuration) is usually diamagnetic.

2. *Geometry of the Complex:* The number of unpaired electrons and the splitting of d-orbitals can be affected by the geometry surrounding the metal. For example, the d-orbitals are split differently in octahedral and tetrahedral environments. If the geometry is known, magnetic susceptibility can reveal details about the electron configuration.

3. *Ligand Field Strength:* Because they contain nitrogen donor atoms, Schiff bases can function as potent field ligands. Depending on the strength of the ligand field, a high-spin to low-spin transition may take place, which would be represented in a shift in magnetic susceptibility.

4. *Determination of Antiferromagnetic Coupling:* There may be magnetic interactions between neighboring metal centers in some dinuclear or polynuclear metal complexes containing Schiff base ligands. Antiferromagnetic coupling may be

indicated if the magnetic moments, which are commonly expressed in terms of Bohr magnetons, or μB, are smaller than one would expect given the number of unpaired electrons.

5. *Verifying Synthesis:* Magnetic susceptibility measurements can help confirm if the desired compound has been synthesized if a targeted metal complex of a Schiff base is expected to have a specific magnetic behavior (based on the metal's oxidation state, coordination number, and geometry).

3.4 Thermogravimetric analysis (TGA) and differential thermal analysis (DTA)

3.4.1 Thermogravimetric analysis (TGA)

TGA calculates how a sample's weight changes in response to temperature or time spent in a controlled environment. Usually, the sample is heated steadily, and any variations in weight are noted.

Decomposition temperature: By identifying the temperature at which Schiff base ligands and their metal complexes begin to break down, TGA can shed light on the thermal stability of these ligands.

Identification of coordinated and crystallized solvents: A lot of metal complexes contain solvent molecules that are either crystallized in the lattice or within the coordination sphere. Based on the temperature and weight loss related to their removal, TGA can detect the presence of these solvents.

Stoichiometry: A metal complex's stoichiometry can be determined using TGA. For example, a complex may indicate the presence of two coordinated or crystallized water molecules if its weight decreases by the amount equivalent to two water molecules.

3.4.2 Differential thermal analysis (DTA)

DTA uses a controlled temperature to measure the difference in temperature between a sample and a reference material, which typically does not go through any phase transitions under the selected conditions.

Phase transitions: Melting points and changes in phase from one crystal to another can be found using DTA. For example, the melting of a Schiff base metal complex might be represented by an endothermic peak on a DTA curve.

Decomposition temperature: Similar to TGA, DTA can be used to ascertain when decomposition begins. On a DTA curve, breakdown events usually show up as exothermic peaks.

Coordination changes: Because these changes may involve energy changes (either endothermic or exothermic), DTA can occasionally be used to detect changes in a metal's coordination sphere, such as ligand dissociation.

3.4.3 Combined TGA-DTA

Many modern instruments offer simultaneous TGA-DTA measurements. This allows researchers to correlate weight loss events from TGA with endothermic or exothermic events from DTA, providing a more comprehensive understanding of the thermal behavior of the sample.

Simultaneous TGA-DTA can provide a complete picture of the thermal events from desolvation and ligand loss to decomposition. For instance, a weight loss in the

TGA curve corresponding to an endothermic peak in the DTA might indicate the loss of a coordinated solvent molecule from a metal complex.

Both TGA and DTA are crucial tools in the characterization of Schiff base ligands and their metal complexes. They provide detailed information about the thermal stability, presence of solvents, phase transitions, and other thermally induced processes in these compounds.

3.5 Electrochemical techniques

Electrochemical techniques are powerful tools to characterize Schiff bases and their metal complexes. They provide valuable information about the redox properties and reactivity of these compounds. Some basic electrochemical techniques include the followings:

3.5.1 Cyclic voltammetry and its role in studying redox features

Cyclic voltammetry (CV) is a powerful electrochemical technique used to study the redox (reduction-oxidation) properties of chemical compounds and materials. It provides valuable information about the electrochemical behavior of species in a solution or on the surface of an electrode. Here is an overview of cyclic voltammetry and its role in studying redox features:

- *Basic Principle:* Cyclic voltammetry involves applying a potential (voltage) to an electrode and measuring the resulting current as the potential is swept back and forth in a repetitive cycle.

- The potential is scanned linearly, starting from a certain initial value and increasing or decreasing at a defined scan rate.

- *Electrochemical Cell:* In a typical setup, a working electrode (often a conductive material, such as a metal or modified carbon electrode), a reference electrode (e.g., Ag/AgCl), and a counter electrode (usually a platinum or carbon electrode) are immersed in an electrolyte solution.

- *Redox Reactions:* Redox reactions occur at the working electrode. When the potential is swept, these reactions lead to the transfer of electrons between the electrode and the species in the solution.

- Oxidation involves the loss of electrons (anodic process), while reduction involves the gain of electrons (cathodic process).

- *Voltammogram:* The primary output of cyclic voltammetry is a voltammogram, which is a plot of current (I) versus applied potential (E). It typically displays distinct peaks corresponding to different redox processes.

- *Information Obtained:* Cyclic voltammetry provides information about the redox potentials (E°) of the species involved in the electrochemical reactions. The position of the peaks in the voltammogram indicates the potentials at which redox processes occur.

- The current response is related to the concentration of electroactive species and the kinetics of the redox reactions.

- *Scan Direction:* Cyclic voltammetry can be performed in two directions: forward (from initial to final potential) and reverse (from final to initial potential). This allows the study of both oxidation and reduction processes.

- *Scan Rate:* The scan rate is an important parameter that can be adjusted. Faster scan rates can reveal kinetic information about the redox reactions, while slower rates can provide more detailed information about the thermodynamics.

- *Applications:* Cyclic voltammetry is widely used in various fields, including chemistry, materials science, electrochemistry, and biochemistry. It is used to study the redox behavior of molecules, determine the stability and reactivity of materials, investigate corrosion processes, and examine the electrochemical behavior of biomolecules, among other applications.

3.5.2 The electrochemical window of Schiff base metal complexes

The electrochemical window of Schiff base metal complexes, such as that of Schiff base compounds, refers to the range of electrochemical potentials (voltages) within which the complex can undergo redox reactions without decomposition or undesired side reactions. These metal complexes consist of a Schiff base ligand coordinated to a metal ion. The electrochemical behavior and window of Schiff base metal complexes can be more complex due to the presence of the metal center, which can participate in redox processes.

The determination of the electrochemical window of Schiff base metal complexes is essential when studying their electrochemical properties, such as their redox behavior and stability. Here are some factors that influence the electrochemical window of Schiff base metal complexes:

- *Nature of the Metal Center:* The choice of metal ions in the complex significantly affects the electrochemical behavior. Different metals have different redox potentials, and the presence of the metal center can introduce additional redox processes. The ligand's coordinating ability to the metal ion also plays a role.

- *Schiff Base Ligand:* The specific structure of the Schiff base ligand and the nature of the substituents on it can impact the electrochemical window. The ligand can donate or accept electrons during redox reactions, affecting the overall electrochemical behavior of the complex.

- *Solvent:* The choice of solvent can influence the electrochemical window. Some solvents can facilitate or hinder redox reactions and affect the stability of the complex.

- *pH:* The pH of the solution can influence the electrochemical behavior, especially for metal complexes with pH-sensitive ligands. Changes in pH can affect the protonation state and reactivity of the complex.

- *Electrode Material:* The type of electrode material used in the electrochemical cell can affect the observed redox potentials. The choice of electrode material can impact the kinetics of the redox reactions.

Determining the electrochemical window of Schiff base metal complexes is crucial for designing and optimizing electrochemical processes involving these compounds, such as in catalysis, sensing, or energy storage applications. Researchers typically use cyclic voltammetry and other electrochemical techniques to investigate the redox properties and electrochemical behavior of Schiff base metal complexes, allowing them to understand the complex interplay of factors that influence the electrochemical window and tailor conditions for specific applications.

Other electrochemical techniques include;

3.5.3 Chronoamperometry

This technique involves applying a constant potential to the sample and monitoring the resulting current over time. It is valuable for studying the electrochemical behavior of Schiff bases and their metal complexes in a time-dependent manner. This can reveal information about reaction mechanisms, stability, and electrode processes.

3.5.4 Electrochemical impedance spectroscopy (EIS)

EIS is used to investigate the electrical properties of Schiff base compounds and their metal complexes at different frequencies. This technique provides insights into factors such as charge transfer resistance, double-layer capacitance, and reaction kinetics.

4. Formation and applications of Schiff base

4.1 Schiff bases as catalytic agents

Schiff base metal complexes are widely used as catalysts [10] in various organic reactions, including oxidation, reduction, and asymmetric synthesis.

4.2 Biological activity

Schiff bases and their metal complexes have shown diverse biological activities, including antimicrobial, antiviral, and anticancer properties. Some Schiff bases are also used as ligands in coordination chemistry with applications in medicinal chemistry.

4.3 Colorimetric sensors

Schiff bases are known for their sensitivity to metal ions. They are used in the development of colorimetric sensors for detecting specific metal ions in solution [3].

4.4 Dyes and pigments

Some Schiff bases exhibit vibrant colors and are used as dyes or pigments in the textile and printing industries.

4.5 Polymer chemistry

Schiff bases can be incorporated into polymers, contributing to the development of materials with specific properties, such as enhanced mechanical strength or conductivity.

4.6 Photonic and electronic devices

Schiff bases and their metal complexes have been investigated for their potential use in photonic and electronic devices due to their optical and electronic properties.

4.7 Antioxidants

Some Schiff bases possess antioxidant properties and are studied for their potential role in preventing oxidative damage.

4.8 Anthelmintic

Anthelmintics serve as therapeutic agents designed to eliminate parasitic worms from infected hosts. The initial efficacious anthelminthic agents, discovered in the twentieth century, incorporated toxic metals like arsenic (atoxyl) or antimony (tartar emetic). These compounds were employed in the treatment of trypanosome and schistosome infestations.

Contemporary anthelminthic agents typically function by inducing paralysis in the parasite, often by impeding muscular contraction, causing damage to the worm to enable elimination by the host's immune system, or modifying parasite metabolism, such as affecting microtubule function. Due to variations in the metabolic requirements among different worm species, drugs highly effective against one type of worm may prove ineffective against others [11].

Worm infestation or helminthiasis stands as a significant health concern, particularly in third-world countries, representing a major global public health problem. Infections caused by parasitic worms and protozoa contribute significantly to global mortality and morbidity. Despite the commercial availability of anthelmintic drugs from diverse chemical groups (e.g., imidazole, benzimidazole, and lactone) for treating worm infections, the prevalence and severity of these diseases are on the rise, partly due to the emergence of multidrug resistance [12].

4.9 Nematicidal activity

The nematicidal activity of Schiff base ligands and their metal complexes has garnered interest in agricultural and environmental research. Schiff bases, known for their versatile coordination properties, can form stable complexes with various metal ions. These complexes have demonstrated potential as nematicidal agents, showing efficacy against nematodes, which are parasitic roundworms that can be detrimental to crops.

The mode of action of these Schiff base metal complexes against nematodes involves interactions with the nematode's biological systems. The metal ions coordinated by Schiff base ligands can disrupt key physiological processes in nematodes, leading to paralysis or death. Additionally, the specific chemical structure of Schiff bases may contribute to their nematicidal activity by targeting essential biological pathways in nematodes.

Research in this area is ongoing, exploring different Schiff base ligands and their metal complexes for their efficacy, selectivity, and environmental impact. The development of nematicidal agents based on Schiff bases holds promise for sustainable and effective strategies in controlling nematode infestations in agriculture.

5. Case studies

The present case study delves into the metal complexes of (E) – N^1- (2 hydroxybenzylidene) nicotinohydrazide Schiff base; synthesis, characterization, and nematicidal activity [1] of Cu(II) and Fe(II) Schiff base ligand complexes; synthesis and characterization [7] marked by its potential significance in the agrochemical sector as it will provide additional chemical pesticides for the prevention and effective control of nematodes. This study explores the intricate chemistry involved in the formation of Schiff base metal complexes and its subsequent applications in chemical pesticides. Schiff base metal complexes have garnered considerable attention due to their diverse applications ranging from catalysis [13, 14] to biological activities. Against this backdrop, our investigation aims to shed light on the unique properties and potential utilities of the synthesized Schiff base metal complex, contributing valuable insights to the ever-expanding field of coordination chemistry. Through a systematic examination of the synthesis route, thorough characterization techniques, and practical applications, this case study seeks to elucidate the compound's structural nuances and explore avenues for its application in cutting-edge scientific endeavors [14].

The synthesis of the Schiff base metal complex involved a novel approach that not only streamlined the process but also resulted in a higher yield compared to traditional methods. This innovative synthesis strategy addresses longstanding challenges in the field and sets the groundwork for more efficient synthetic routes in the future.

Advanced spectroscopic techniques, including TGA and crystallography, provided detailed structural insights into the metal complex. Of particular significance is the C=N bond, which plays a pivotal role in its influence on reactivity, stability, etc. These revelations mark a substantial contribution to the understanding of Schiff base metal complex structures.

Metal complexes containing Mo(IV) and Mn(II) paired with ligands hydrazine carboxamide and hydrazine carbothiamide exhibit notable antibacterial effects against *S. aureus* and *Xanthomonas campestris.* Additionally, antibacterial activities are observed in tridentate Schiff bases and their associated metal complexes against *E. coli*, *S. aureus, B. subtilis*, and *B. pumilis*, as documented in reference [15]. Azan, *et al.* [16] reported on the antimicrobial potential of Schiff bases derived from naphtha-[1, 2-d]-thiazol-2-amine and their corresponding metal complexes in reference [17].

Schiff bases demonstrate anti-inflammatory properties, act as inhibitors for allergic responses, exhibit activity in reducing radicals, and have analgesic effects [10]. Thiazole-based chitosan and carboxymethyl-chitosan display antioxidant capabilities, including scavenging of superoxide and hydroxyl radicals [18]. Metal complexes of furan semicarbazone show noteworthy anthelmintic and analgesic activities [19].

Transition metal complexes involving Cu, Zn, and Cd hinder the cellular proliferation of MDA-MB-231 breast cancer cells. Among these complexes, $Cd(C_{18}H_{16}N_3O_2)_2.2CH_3OH$ exhibits the most potent antiproliferative activity [20]. In a separate study, Shabani *et al.* [21] have documented the anti-tumor activity of Iron(III) Schiff base complexes.

Schiff bases, renowned for their diverse biological functions, have undergone extensive investigation regarding their coordination tendencies with metals, leading to the development of highly efficient catalysts [22, 23]. These complexes, especially those involving transition metals, exhibit remarkable catalytic efficacy across various reactions, even in the presence of moisture and at elevated temperatures [24]. The notable thermal and moisture stability of these complexes enhances their utility in catalytic applications, and ongoing research is exploring their potential in the development of novel chemotherapeutic agents [22, 23].

A catalytically active Mn(II) Schiff base complex, featuring a triazole structure, has been identified as an efficient catalyst for the Henry reaction, yielding b-hydroxyl nitroalkanes in high yield [14]. Additionally, Co(II), Fe(III), and Ru(III) complexes derived from Schiff bases, synthesized from hydroxyl benzaldehyde, have demonstrated effective catalysis in the oxidation of cyclohexane to cyclohexanol and cyclohexanone in the presence of hydrogen peroxide [25].

6. Conclusion

In conclusion, the thorough examination of Schiff base ligands and their metal complexes has yielded significant understandings into the fascinating field of coordination chemistry. By thoroughly examining their spectroscopic, chemical, and physical characteristics, we have been able to comprehend these compounds' coordination modes and structural characteristics on a deeper level.

The utilization of various spectroscopic techniques, including X-ray crystallography, NMR, IR, and UV-visible, has been beneficial in clarifying the molecular structures of the ligands and their metal complexes. These analyses have clarified the coordination environment surrounding the metal centers in addition to confirming the successful formation of the intended compounds.

Additionally, the investigation has brought attention to the effects of various metal ions on the characteristics of Schiff base ligands, underscoring the significance of metal-ligand interactions in controlling the general behavior of these complexes. We now have a better understanding of the variables affecting the stability and reactivity of the complexes thanks to the investigation of the electronic and steric effects of the metal ions.

The importance of this work is highlighted by the potential uses of Schiff base ligands and their metal complexes in a variety of industries such as material science, medicinal chemistry, and catalysis. The information gathered from this research not only broadens our understanding of these compounds but also opens up new avenues for further study and application within the scientific community.

To sum up, the identification of Schiff base ligands and their metal complexes is an important step toward improving our understanding of coordination chemistry. The diverse characteristics of these substances present an abundant opportunity for additional investigation and utilization, rendering them auspicious contenders for an extensive array of scientific and technological pursuits.

Author details

Mohammed Umar Adaji[1*], Moses Saviour Iorungwa[2] and Olalekan Wasiu Salawu[3]

1 Kogi State University, Anyigba, Nigeria

2 Jodeph Sarwuan Tarker University, Makurdi, Nigeria

3 Federal University, Lokoja, Nigeria

*Address all correspondence to: adajimed@gmail.com; adaji.mu@ksu.edu.ng

References

[1] Adaji MU, Wuana RA, Itodo AU, Eneji IS, Iorungwa MS. Metal complexes of (E) – N1- (2 -hydroxybenzylidene) nicotinohydrazide Schiff base; Synthesis, characterization and nematicidal activity. Arabian Journal of Chemical and Environmental Research. 2022;**9**(1):1-27

[2] Aminu IH, Aliyu HN, Musa H, Dayyib AS. Synthesis, Characterization and antimicrobial studies on Mn(II), Fe(II), Co(II) complexes of Schiff base derived from 3-Formylchromone and Benzohydrazide. *Algerian Journal of*. Engineering and Technology. 2021;**4**:81-89. Available from: http://jetjournal.org/index.php/ajet/article/view/45

[3] Maity S, Shyamal M, Das D, Mazumdar P, Sahoo GP, Misra A. Aggregation induced emission enhancement from antipyrine-based Schiff base and its selective sensing towards picric acid. Sensors and Actuators B: Chemical. 2017;**248**:223-233. DOI: 10.1016/j.snb.2017.03.161

[4] Pawlica D, Marszałek M, Mynarczuk G, Sieroń L, Eilmes J. New unsymmetrical Schiff base Ni(II) complexes as scaffolds for dendritic and amino acid superstructures. New Journal of Chemistry. 2004;**28**:1615-1621

[5] Ahmad N, Alam M, Wahab R, Ahmed M, Ahmad A. Synthesis, spectral and thermo-kinetics explorations of Schiff-base derived metal complexes. Open Chemistry. 2020;**18**(1):1304-1315. DOI: 10.1515/chem-2020-0168

[6] Salawu OW, Iorungwa MS, Adaji MU. Synthesis, characterization and kinetic studies of Fe(II) and Cu(II) complexes of nicotinic acid hydrazide. FUW Trends in Science and Technology Journal. 2016;**1**(2):534-538

[7] Adaji MU, Wuana RA, Itodo AU, Eneji IS, Iorungwa MS. Nematicidal activity of Cu(II) and Fe(II) Schiff base ligand complexes: Synthesis and characterisation. FUAM Journal of Pure and Applied Science. 2021;**1**(1):99-109

[8] Mirza AH, Hamid SA, Malai H, Aripin S, Karim MR, Arifuzaman M, et al. Synthesis, spectroscopy and X-ray crystal structures of some zinc(II) and cadmium(II) complexes of the 2-pyridinecarboxaldehyde Schiff bases of S-methyl- and S-benzyldithiocarbazates. Polyhedron. 2014;**74**:16-23. DOI: 10.1016/j.poly.2014.02.016

[9] Salawu OW, Iorungwa MS, Adaji MU. Synthesis, characterization, kinetic and thermodynamic studies of Co (II) and Zn (II) complexes of nicotinic acid hydrazide. Islamic University Multidisciplinary Journal. 2019;**6**(2):281-288

[10] Kajal A, Bala S, Sunil K, Sharma N, Saini V. Schiff bases: A versatile pharmacophore. Journal of Catalyst. 2013;**2013**:1-14

[11] Chai JY, Jung BK, Hong SJ. Albendazole and mebendazole as anti-parasitic and anti-cancer agents: An update. Korean Journal of Parasitology. 2021;**59**(3):189-225

[12] Cooper KM, Whelan M, Kennedy DG, Trigueros G, Cannavan A, Boon PE, et al. Anthelmintic drug residues in beef: UPLC-MS/MS Method validation, European retail beef survey, and associated exposure and risk assessments. Food Additives and Contaminants Part A. 2012;**29**(5):746-760. DOI: 10.1080/19440049.2011.653696

[13] Gupta KC, Sutar AK. Catalytic activities of Schiff base transition metal

complexes. Coordination Chemistry Review. 2008;**252**(12-14):1420-1450. DOI: 10.1016/j.ccr.2007.09.005

[14] Alshaheri AA, Tahir MIM, Abdulrahman MB, Tawfika TB. Synthesis, characterisation and catalytic activity of dithiocarbazate Schiff base complexes in oxidation of cyclohexane. Journal of Molecular Liquids. 2017;**240**:486-496. DOI: 10.1016/j.bmc.2006.01.054

[15] Chatterjee DM, Shepherd RE. Oxo-transfer catalysis from t-BuOOH with C-H bond insertion using tridentate Schiff-base-chelate complexes of ruthenium(III). Inorganica Chimica Acta. 2004;**357**(4):980-990. DOI: 10.1016/j.ica.2003.07.008

[16] Chohan ZH, Scozzafava A, Supuran CT. Zinc complexes of benzothiazole-derived Schiff bases with antibacterial activity. Journal of Enzyme Inhibition and Medicinal Chemistry. 2003;**18**(3):259-263. DOI: 10.1080/1475636031000071817

[17] Azan F, Singh S, Sukhbir LK, Prakash O. Synthesis of Schiff bases of naphtha[1,2 d]thiazol-2-amine a nd metal complexes of 2-(2′-hydroxy) benzylideneaminonaphthothiazole as potential antimicrobial agents. Journal of Zhejiang University Science. 2007;**8**:446-452. DOI: 10.1631/jzus.2007.B0446

[18] Desai SB, Desai PB, Desai KR. Synthesis of some Schiff bases, thiazolidones, and azetidinones derived from 2,6-diaminobenzo[1,2-*d*:4,5-*d′*]bisthiazole and their anticancer activities. Heterocycling Communication. 2001;7(1):83-90. DOI: 10.1515/HC.2001.7.1.83

[19] Sathe BS, Jaychandran E, Jagtap VA, Sreenivasa GM. Synthesis characterization and anti-inflammatory evaluation of new fluorobenzothiazole Schiff's bases. International Journal of Pharmaceutical Research and Development. 2011;**3**(3):164-169. DOI: 10.1155/2013/893512

[20] Sondhi SM, Singh N, Kumar A, Lozach O, Meijer L. Synthesis, anti-inflammatory, analgesic and kinase (CDK-1, CDK-5 and GSK-3) inhibition activity evaluation of benzimidazole/benzoxazole derivatives and some Schiff's bases. Bioorganic and Medicinal Chemistry. 2006;**14**(11):3758-3765. DOI: 10.1016/j.bmc.2006.01.054

[21] Pandey A, Dewangan D, Verma S, Mishra A, Dubey RD. Synthesis of Schiff bases of 2-amino-5-aryl-1,3,4-thiadiazole and its analgesic, anti-inflammatory, anti-bacterial and anti-tubercular activity. International Journal of ChemTech Research. 2011;**3**(1):178-184. DOI: 10.1155/2012/145028

[22] Shabani F, Saghatforoush LA, Ghammamy S. Synthesis, characterization and anti-tumour activity of iron (iii) Schiff base complexes with unsymmetric tetradentate ligands. Bulletin of Chemical Society of Ethiopia. 2010;**24**(2):193-199. DOI: 10.4314/bcse.v24i2.54741

[23] Abu-Dief AM, Ibrahim M, Mohamed A. A review on versatile applications of transition metal complexes incorporating Schiff bases. Beni-Suef University Journal of Basic and Applied Sciences. 2015;**4**(2):119-133. DOI: 10.1016/j.bjbas.2015.05.004

[24] Dalia SA, Afsan F, Khan MN, Zakaria MC, Zahan Kudirat E. A short review on chemistry of Schiff base metal complexes and their catalytic application. International Journal of Chemical Studies. 2018;**6**(3):2859-2866

[25] Nesterova O, Nesterov DS, Krogul-Sobczak A, Fátima MC, da Silva G, Pombeiro A. Synthesis, crystal structures and catalytic activity of Cu(II) and Mn(III) Schiff base complexes: Influence of additives on the oxidation catalysis of cyclohexane and 1-phenylehanol. Journal of Molecular Catalysis A: Chemical. 2017;**426**(Part B):506-515

Chapter 8

Schiff Base Moiety Could be a Possible Inhibitor of Mutated Nrf2/KEAP1 Associated Cancer

Malavalli Guruswamy Dileep Kumar and Jayarama Shankar

Abstract

In recent studies, the cap'n'collar (CNC) bZIP transcription factor Nrf2 plays a pivotal role in cancer drug development. The Nrf2/KEAP1 pathway is the most important signalling cascade involved in the resistance of oxidative damage induced by external chemicals. The Nrf2 maintains cellular homeostasis, anti-inflammatory, and anticancer properties by activating downstream signalling pathways and their cell survival. But, a recent literature survey suggested that mutated KEAP1/Nrf2 is responsible for cancer formation by suppressing apoptosis and metabolic reprogramming. More importantly, Nrf2 is proven to contribute to the chemoresistance and radioresistance of cancer cells as well as inflammation-induced carcinogenesis. A number of Nrf2 inhibitors, particularly Schiff base molecules discovered for cancer treatment, were reviewed in this chapter. Schiff bases or azomethines are compounds formed by a condensation reaction between primary amines and aldehydes, and have various biological, medicinal, clinical, pharmacological and analytical applications. These provide a new strategy that targeting Nrf2 could be a promising therapeutic approach against cancer. This review emphasises the role of Schiff base to summarise the effects of Nrf2 in cancer, revealing its function both in cancer prevention and inhibition, to further synthesise the novel Schiff base-related anticancer treatment.

Keywords: schiff base, Nrf2, KEAP1, oxidative stress, cancer, inflammation, chemotherapy

1. Introduction

Cancer is a multistage progressive disease in which a group of cells start replicating uncontrolled after a significant transformation occurs in hereditary modifications of normal cells [1–5]. Genes regulating reproduction, cellular growth and cell cycle get mutated and convert normal cells into cancerous cells [6–8]. Such cells escape cell cycle checkpoints, evade apoptosis, produce bigger populations, increase their mass, and start invasion into nearby tissues [9, 10]. Cancer affects almost all body systems, organs and tissues, except dead cells like hair and nails, and is continuously detected in individuals of all ages. After affecting a specific body part or system, the disease also progresses to other areas of the body. Cancer can be managed and treated effectively if diagnosed early, but it can be fatal if it remains undiagnosed and the disease progresses to an end-stage [11–14]. There are more than 270 types of cancer that have a great tendency to resist chemotherapeutics, leading to a relapse after an initial cure. Therefore, the discovery of novel molecules and strategies to combat this challenge is necessary [15]. Apoptosis is a programmed cell death which prevents abnormal cells from further replication. In cancer cells, apoptosis is blocked by activating anti-apoptotic proteins that lead to the over-activation of cancer cells to grow further. Hence, activation of apoptosis in cancer cells is a significant way to develop anticancer drugs that trigger the production of pro-apoptotic enzymes and inhibit downstream signalling pathways. Thus, Schiff bases could be one of the promising drug targets for many cancers in humans [16]. Schiff bases or azomethines are compounds formed by a condensation reaction between primary amines and aldehydes and have various biological, medicinal, clinical, pharmacological and analytical applications [17–20]. Aromatic primary amines can also contain additional donor functional groups like –Cl, –OH, –CH3, etc., which help enhance and regulate biological activities [21–23]. Schiff bases have been shown to exhibit various biological activities that include anticancer and anti-inflammatory [24–26].

The formation of tumours and its malignancy is tightly controlled by the external exposure of cells to harmful or protective environments. The possible link between these mechanisms is seen at the molecular level via modulation of the signalling cascade for which the initiation or inhibition can either sustain or impair cell proliferation of normal cells and increase the progression of cancer cells. The Nrf2 is involved in both normal and cancer pathways. In cancer, the Nrf2 signalling pathway is the most important for cell protection from oxidative stress and electrophilic effects [27]. The NFE2L2 quality encodes the Nrf2 protein, which is essential for the "Cap 'n' Collar" (CNC) group of transcriptional factors (TFs).

2. Canonical and non-canonical pathway of Nrf2

The reactive oxygen species level is the key element in the Nrf2 and KEAP1 pathways. In most of the KEAP1/Nrf2 pathway, KEAP1 undergoes point mutation, and KEAP1 helps Nrf2 release from the inhibitor and translocates into the nucleus. KEAP1 protein family belongs to the Kelch family of proteins (KLHL); KEAP1 interacts with the Nrf2-ECH homology (Neh2) domain of Nrf2 in a specific amino acid sequence containing 69–84 and is flanked by the ETGE motif [28–30]. The cullin-3 is a ubiquitination ligase, and it has a direct link with the Nrf2 proteins. The KEAP1 is responsible for the continuous degradation of lysine in the Heh2 domain and its proteasomal degradation; this evidence clearly indicates that there is a possible link between Nrf2 and its cullin-3. The epigenetic regulation in the KEAP1 leads to the over-activation of Nrf2, and the adaptor proteins called p62 are involved in the over-activation of Nrf2. The p62 is the key mediator and the target for the Nrf2 transcriptional factor and thus increases the overstimulation of Nrf2 for cell proliferation. The P21 also plays a pivotal role in the activation of Nrf2 because it has a positive regulatory effect on the Nrf2 protein. Some of the cytokines like IL-1β and TNF-α also triggered due to the production of adiponectin in adipocyte-associated macrophages, and it triggers the production of p62, a negative regulator of pro-inflammatory cytokines, p62 interacts with KEAP1 and mediates its degradation (**Figure 1**) [31].

The primary purposes for the over articulation level of Nrf2 in malignant growth are DNA mutation, epigenetic modifications, and alteration in the protein

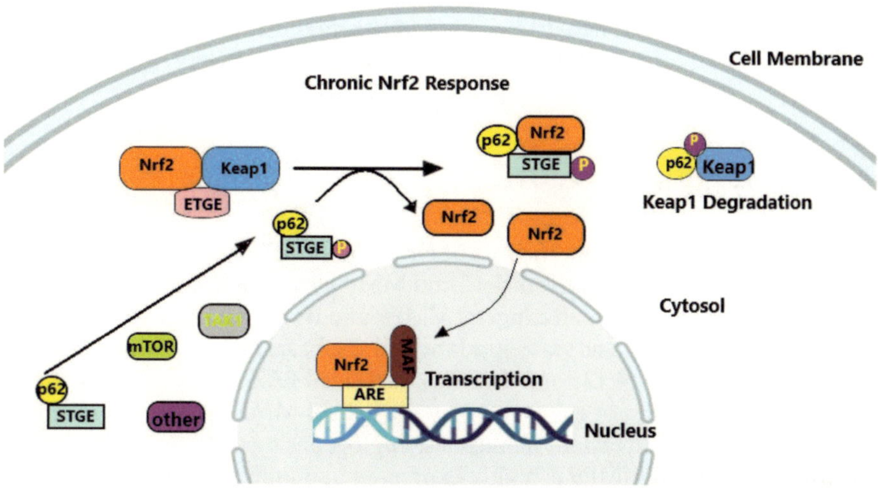

Figure 1.
The role of adaptor protein p62 in Nrf2 signalling pathway – The phosphorylation of P62 by mTOR, and TAK1 leads to the stronger binding affinity towards KEAP1 due to reassemble of ETGE in Nrf2. As a result, Nrf2 translocates into the nucleus and binds on the ARE region along with MAF and finally triggers transcription.

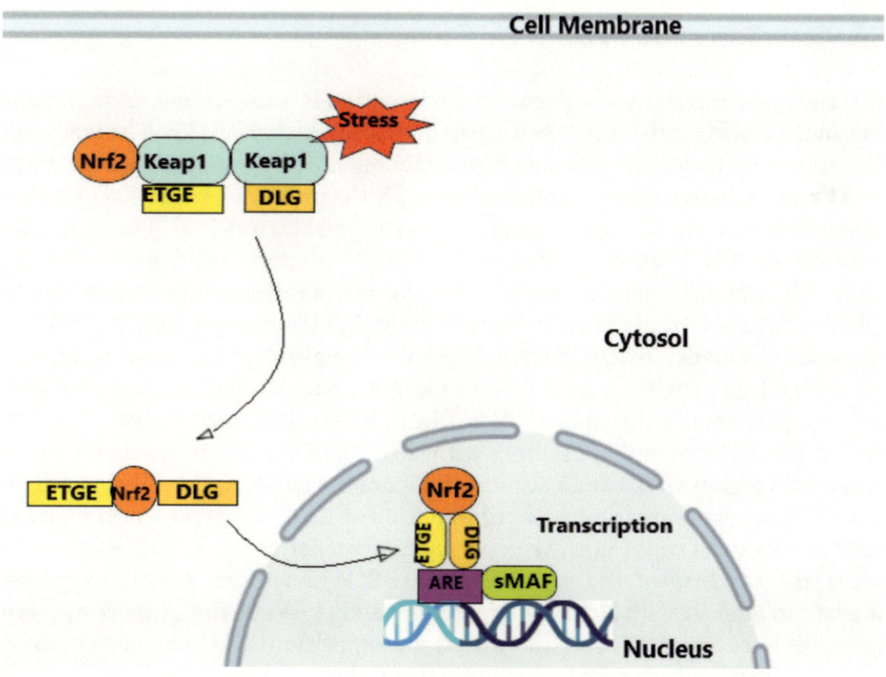

Figure 2.
Mechanistic approach of KEAP1/Nrf2 in cancer – Due to the modification of KEAP1 cysteine leads to the release of Nrf2, ETGE and DLG. The bunch of Nrf2, ETGE, and DLG enters into the nucleus and interacts with ARE promoter region on the DNA along with sMAF that leads to the over activations of transcription in mutated Nrf2 cells.

structure [32]. In some disease types, one or both KEAP1 and Nrf2 are changed, which does not permit legitimate chemical modifications between both Nrf2 as well as KEAP1. In most cancers, the regular methylation pattern of the DNA is altered, and the promoter region of KEAP1 becomes hyper methylated, resulting in the low level of transcription of the KEAP1 protein and the release of Nrf2 in specific cells (**Figure 2**).

3. Liver cancer and Nrf2

In hepatocellular carcinoma, the role of Nrf2 and MMP9 was well studied, and there is a direct correlation between Nrf2 and MMP9 levels in the activation of transcription factors in many human cancers. MMP9 and its role in angiogenesis are crucial in invasion and metastasis via degradation of the basal membrane and Bcl family protein. In most cancers, Nrf2 contributes to decreased BAX and over-simulations of Bcl2 [33]. The degradation of collagen IV by the help of MMP9 is responsible for the cancer cell progression, invasion, metastasis, angiogenesis and their proliferation. Apart from liver cancer, MMP9 is highly expressed in mammary carcinoma cell lines. The Gelatinase B plays a pivotal role in the progression of cancer cells via angiogenesis to stromal remodelling and metastasis. In Glioma cells, levels of MMP9 and Nrf2 are highly expressed. Once the ARE1 promoter regions are more specific to the mutated Nrf2 and MMP9 (due to KEAP1 mutation), the bunch of transcriptional factors along with Nrf2 translocates into the nucleus and activates transcription [34].

4. Cervical cancer and Nrf2

Cervical cancer is the main global health issue and it is the second most diagnosed cancer in the world, with 5 lakhs new cases diagnosed each year [35]. The main role of Nrf2 is to control the redox potential by controlling the physiological balance in the body. The antioxidant response element (ARE) is the promoter where Nrf2 initiates its first mode of action along with other transcriptional factors [36]. The *antioxidant response element* (ARE) contains the Leucine zipper structure, and it belongs to Cap n collar family, and its transactivation induces the gene expressions to tolerate oxidative damage and electrophilic effect. In a normal physiological process, the Nrf2 is expressed at a moderate level because the KEAP1 regulates Nrf2 in a controlled manner [37]. Due to oxidative stress or a xenobiotic stress, KEAP1 is degraded and cysteine residue gets modified in KEAP1. Due to its modification of cysteine residue in the KEAP1, Nrf2 forms heterodimer with sMAF family protein and finally interacts with the ARE region on the DNA site and activates the transcription. Hence, KEAP1 acts as a tumour suppressor role; loss of Mutation of KEAP1 regulates over expression of Nrf2 in the cancer cells [38]. *The hyper methylation pattern of KEAP1 in cervical cancer is poorly understood*. Breast cancer gene (BRCA) is another tumour suppressor gene that can interact with Nrf2 to increase its stability. When BRCA is absent, oestrogen restores Nrf2 activation and diminishes the expression of ROS in many mammary carcinomas. Hence, our future goal is more focused towards epigenetic regulation of KEAP1 in the cervical cancer and their inhibiting drugs, particularly Schiff bases, are more preferable [39].

5. Lung cancer and Nrf2

The hinge and latch model explains the direct indication of Nrf2/KEAP1 regulation in ETGE and DLG [40]. During this interaction, the Nrf2/KEAP1 protein adopts conformational changes with the KEAP1 molecule. In cancer cells, the redox potential is completely imbalanced, and it exhibits high ROS production due to oncogenic production by KEAP1 modification. In this contest, Nrf2 plays a significant role in both tumour suppressor as well as oncogenic roles. Both canonical and non-canonical pathways are key pathways to study about Nrf2 [41]. In the canonical pathway, the Nrf2 are low expressive and have a low affinity towards cul-3/RBX1 E3 ubiquitin-dependent degradation [42]. During its conformational change in Nrf2/KEAP1, newly formed Nrf2 interacts with ETGE to alternative subunits of KEAP1 and finally interacts with DLG for its closed conformation and its final degradation from 26 s proteasome complex (**Figure 3**).

In non-canonical pathways, Nrf2 is independently regulated by KEAP1. In this manner, the ubiquitin-binding autophagy receptor p62/sequestosome-1 (p62/SQSTM1) is a positive regulator of ARE in an independent redox state. The P62/SQSTM1 binds with KEAP1 through its STGE motif and promotes further degradation to increase the level of Nrf2 protein expression [43]. The glycogen synthase kinase 3 beta (GSK-3β) is another positive regulator of Nrf2. The GSK-3β can directly involve phosphorylates β-TrCP to ubiquitination of Nrf2. It also phosphorylates Fyn, and thus in turn activate Nrf2 and it helps to return to cytosol, and further

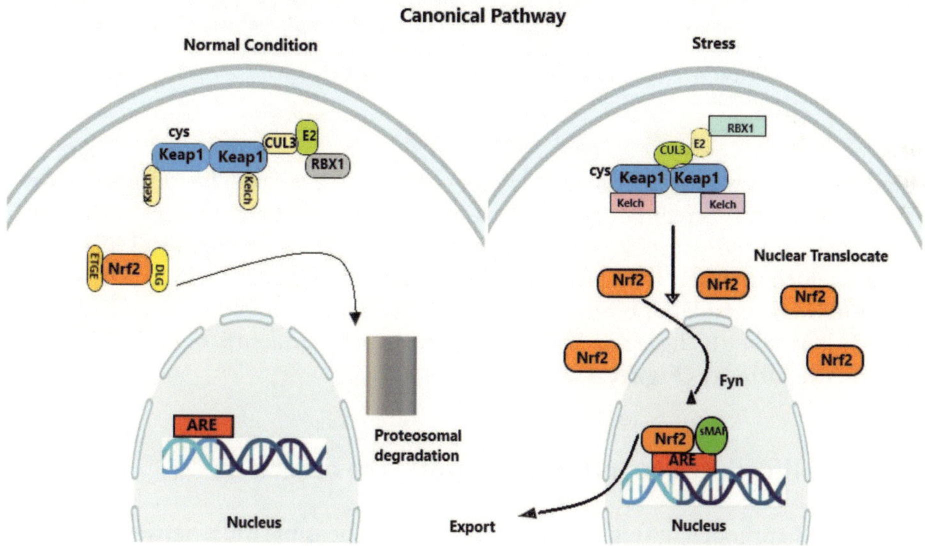

Figure 3.
Canonical Pathway of Nrf2- Under normal conditions, Nrf2 ETGE/DLG motifs interact with the KEAP1 Kelch domain. After this interaction with KEAP1, CUL3/RBX1 E3 interacts with KEAP1 and then targets Nrf2 degradation through 26 s proteasome complex. Due to oxidative and electrophilic attacks, cysteine residues get altered in KEAP1 protein leading to the disruption of Nrf2/KEAP1 interaction. After the disruption of Nrf2 with KEAP1, Nrf2 accumulation in the cytosol increases and then translocates into the nucleus to form heterodimer with sMAF and activates transcription. After translocation of Nrf2, Nuclear export of Nrf2 is possible due to phosphorylation in Src family proteins such as Fyn.

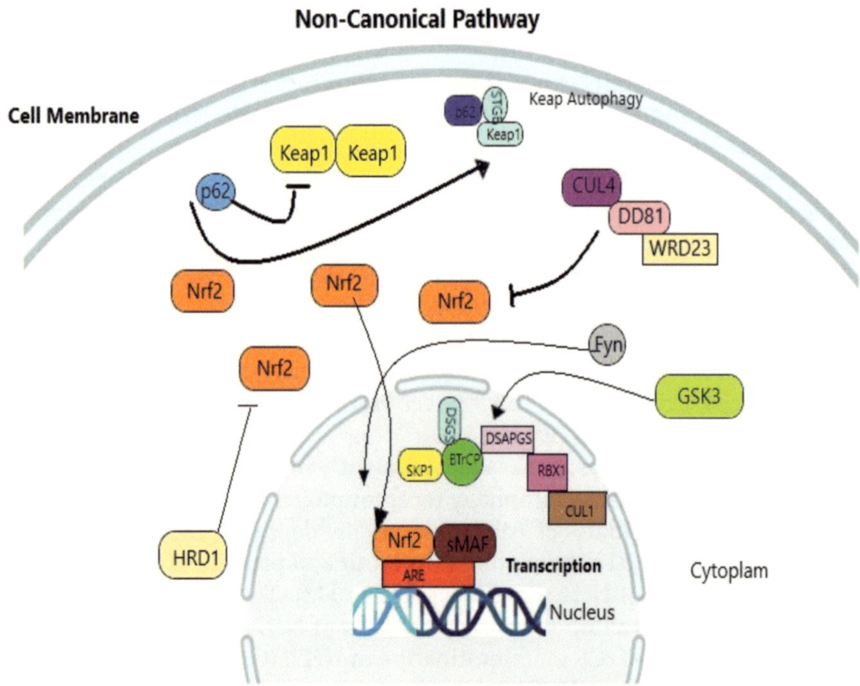

Figure 4.
In non-canonical pathways, Nrf2 can independently activate by p62/SQSTM1 which further activates the KEAP1 autophagy degradation through STGE binding motifs. The CUL3/SKP1 complex initiates the degradation and ubiquitination of Nrf2. In ER stress pathway, HRD1 interacts with Neh4 and 5 domains and activates Nrf2. In a recent review, CUL/DDB1/WDR23 was identified as another form of E3 ubiquitin ligase to activate and control Nrf2, and its mode of action is still not well understood.

degradation takes place via proteasomal degradation in cytosol. The PI3/AKT pathway is also involved in this regulatory mechanism due to its ability to interact with GSK-3β [44]. HMG-CoA reductase degradation protein 1 (HRD1) is another Nrf2 E3 ubiquitin ligase which interacts with Nrf2 and causes further degradation in ER stress [45]. The cullin4/damaged DNA binding protein-1/WD Repeat Domain 23 (CUL4/DDB1/WDR23) is another E3 ligase of Nrf2 that interacts independently of cul3/KEAP1 in Nrf2/KEAP1 pathway (**Figure 4**).

6. Possible link between inflammation and the cancer

Our novel approach in this chapter is to study the possible links between inflammation and cancer through Nrf2 signalling cascade by activation of MAPK in both cancer and inflammation-related pathways. The cytoprotective and antioxidant expressions are regulated by Nrf2, and they protect from electrophilic and oxidative stress and inhibit the accumulations of pro-inflammatory mediators, thereby contributing to the basic way to prevent inflammation-associated cancer [46]. Based on the requirement, our project is more interested in designing new novel molecules that are basically inhibitors to suppress the Nrf2 and KEAP1 mediated pathways that lead to the over-activation of cancer for their cellular growth and metastasis.

7. Schiff base and Nrf2

Novel Schiff base molecules are found to have a wide range of pharmacological activities and attract the researchers to work on this nucleus. Schiff base and its targets with Nrf2 are not well studied; hence we are more focused towards molecular interaction of Schiff base with KEAP1/Nrf2 pathway. Literature survey indicates that many novel Schiff base molecules have been synthesised with the aim of getting compounds of significant anti-inflammatory and anticancer activities with reduced adverse effects [47]. This chapter aims to compile the reports on novel Schiff base molecules possessing anti-inflammatory and anticancer activities. This chapter also includes the reports on Schiff base of existing anticancer medication in future days.

8. Chemistry and scheme of the proposed molecule for this project (d1) and (d2)

Based on the above observation, our group has designed Schiff base molecules to target the KEAP1 and Nrf2 pathway, and we are planning to work with this project in upcoming days. For the syntheses of compounds (d1) and (d2), to a solution of 2-methoxyaniline (100 mg, 0.813 mmol) in ethanol (15 ml) an equimolar amount of 5-bromo-2-hydroxy-3- methoxybenzaldehyde (187 mg, 0.813 mmol) for (d1) or 2-hydroxy-3- methoxybenzaldehyde (123 mg, 0.813 mmol) for (d2), was added dropwise with constant stirring, in the presence of a catalytic amount of glacial acetic acid. These solutions were heated under reflux for 4 h, after which the reactions were judged from TLC to be complete [48]. The mixtures were cooled to ambient temperature, and the resulting solid products were collected by filtration and then recrystallised from N, N-dimethylformamide for (d1) and from acetonitrile for (d1 and d2).

9. Conclusion

Attributable to their great pharmacological activities, Schiff base and their analogs are fundamental and essential drugs for disease treatment, particularly those wherein oxidative stress and inflammation have been involved. This is because of the way that the majority of these Schiff bases enact the NRF2 signalling pathway, which directs hindrance of Nrf2 and KEAP1, the critical go-between in the advancement of chemotherapeutic medications. Through their NRF2-intervened cell reinforcement and mitigating impacts, Schiff base constricts the slow decrease in disease related to Nrf2 and KEAP1 mutation. Besides, the ability of these Schiff bases to restrain the articulation levels of NRF2, for example, DLG and ETGE addresses a fundamental restorative technique for human disease. In rundown, in view of NRF2-interceded activities, pharmacological profile and SAR assessment, Schiff base and their analogs address great possibility for additional advancement for anticancer medication.

Acknowledgements

Our studies and preparations of this chapter were supported by the Food technology Department, Davangere University, India.

Conflict of interest

There is no conflict of interest.

Author details

Malavalli Guruswamy Dileep Kumar[1] and Jayarama Shankar[2]*

1 PG Department of Biotechnology, Teresian College, Karnataka, India

2 PG Department of Food Technology, Davangere University, Karnataka, India

*Address all correspondence to: sankkar.bio@gmail.com

References

[1] Ruddon RW. Cancer Biology. Oxford University Press; 2007. pp. 1-553

[2] Guruswamy DKM, Balaji KDS, Dharmappa KK, Jayarama S. Novel 3-(3, 5-difluoro-4-hydroxyphenyl)-1-(naphthalen-2-yl) prop-2-en-1-one as a potent inhibitor of MAP-kinase in HeLa cell lines and anti-angiogenic activity is mediated by HIF-1α in EAC animal model. Oncotarget. 2020;**11**(50):4661

[3] Narasimhamurthy KH, Guruswamy DKM, Kallesha N, Rangappa KS. Synthesis of bioactive quinoline acting as anticancer agents and their mode of action using in silico analysis towards Aurora kinase A inhibitors. Chemical Data Collections. 2021;**35**:100768

[4] Guruswamy DKM, Jayarama S. Proapoptotic and anti-angiogenic activity of (2E)-3-(2-bromo-6-hydroxy-4-methoxyphenyl)-1-(naphthalene-2-yl) prop-2-en-1-one in MCF7 cell line. Chemical Papers. 2020;**74**(7):2229-2237

[5] Raveesha R, Anusuya AM, Raghu AV, Kumar KY, Kumar MD, Prasad SB, et al. Synthesis and characterization of novel thiazole derivatives as potential anticancer agents: Molecular docking and DFT studies. Computational Toxicology. 2022;**21**:100202

[6] Levine AJ, Perry ME, Chang A, Silver A, Dittmer D, Wu M, et al. The 1993 walter hubert lecture: The role of the p53 tumour-suppressor gene in tumorigenesis. British Journal of Cancer. 1994;**69**(3):409-416

[7] Fu J, Liao L, Balaji KS, Wei C, Kim J, Peng J. Epigenetic modification and a role for the E3 ligase RNF40 in cancer development and metastasis. Oncogene. 2021;**40**(3):465-474

[8] Lingaraju GS, Balaji KS, Jayarama S, Anil SM, Kiran KR, Sadashiva MP. Synthesis of new coumarin tethered isoxazolines as potential anticancer agents. Bioorganic & Medicinal Chemistry Letters. 2018;**28**(23-24):3606-3612

[9] Roopashree R, Mohan CD, Swaroop TR, Jagadish S, Raghava B, Balaji KS, et al. Novel synthetic bisbenzimidazole that targets angiogenesis in Ehrlich ascites carcinoma bearing mice. Bioorganic & Medicinal Chemistry Letters. 2015;**25**(12):2589-2593

[10] Weinberg RA. How cancer arises. Scientific American. 1996;**275**(3):62-70

[11] Sporn MB, Liby KT. NRF2 and cancer: The good, the bad and the importance of context. Nature Reviews Cancer. 2012;**12**(8):564-571

[12] Lau A, Villeneuve NF, Sun Z, Wong PK, Zhang DD. Dual roles of Nrf2 in cancer. Pharmacological Research. 2008;**58**(5-6):262-270

[13] de la Vega MR, Chapman E, Zhang DD. NRF2 and the hallmarks of cancer. Cancer Cell. 2018;**34**(1):21-43

[14] Menegon S, Columbano A, Giordano S. The dual roles of NRF2 in cancer. Trends in Molecular Medicine. 2016;**22**(7):578-593

[15] Wu S, Lu H, Bai Y. Nrf2 in cancers: A double-edged sword. Cancer Medicine. 2019;**8**(5):2252-2267

[16] Bailar JC, Gornik HL. Cancer undefeated. New England Journal of Medicine. 1997;**336**(22):1569-1574

[17] Ebosie NP, Ogwuegbu MO, Onyedika GO, Onwumere FC. Biological and analytical applications of Schiff

base metal complexes derived from salicylidene-4-aminoantipyrine and its derivatives: A review. Journal of the Iranian Chemical Society. 2021;**18**(12):3145-3175

[18] Ashraf MA, Mahmood K, Wajid A, Maah MJ, Yusoff I. Synthesis, characterization and biological activity of Schiff bases. IPCBEE. 2011;**10**(1):185

[19] Hameed A, Al-Rashida M, Uroos M, Abid Ali S, Khan KM. Schiff bases in medicinal chemistry: A patent review (2010-2015). Expert Opinion on Therapeutic Patents. 2017;**27**(1):63-79

[20] Arulmurugan S, Kavitha HP, Venkatraman BR. Biological activities of Schiff base and its complexes: A review. Rasayan Journal of Chemistry. 2010;**3**(3):385-410

[21] Ahlberg E, Amberg A, Beilke LD, Bower D, Cross KP, Custer L, et al. Extending (Q) SARs to incorporate proprietary knowledge for regulatory purposes: A case study using aromatic amine mutagenicity. Regulatory Toxicology and Pharmacology. 2016;**77**:1-12

[22] Gidwani B, Sahu V, Shukla SS, Pandey R, Joshi V, Jain VK, et al. Quantum dots: Prospectives, toxicity, advances and applications. Journal of Drug Delivery Science and Technology. 2021;**61**:102308

[23] Paul S, Pan S, Mukherjee A, De P. Nitric oxide releasing delivery platforms: Design, detection, biomedical applications, and future possibilities. Molecular Pharmaceutics. 2021;**18**(9):3181-3205

[24] Abu-Dief AM, Mohamed IM. A review on versatile applications of transition metal complexes incorporating Schiff bases. Beni-suef University Journal of Basic and Applied Sciences. 2015;**4**(2):119-133

[25] Kajal A, Bala S, Kamboj S, Sharma N, Saini V. Schiff bases: A versatile pharmacophore. Journal of Catalysts. Hindawi Publishing Corporation. 2013;**2013**:14. Article ID 893512. DOI: 10.1155/2013/893512

[26] Nath BD, Islam MM, Karim MR, Rahman S, Shaikh MAA, Georghiou PE, et al. Recent progress in metal-incorporated acyclic schiff-base derivatives: Biological aspects. ChemistrySelect. 2022;**7**(14):e202104290

[27] Zhang DD. Mechanistic studies of the Nrf2-Keap1 signaling pathway. Drug Metabolism Reviews. 2006;**38**(4):769-789

[28] Zhao H, Wang Y, Liu Y, Yin K, Wang D, Li B, et al. ROS-induced hepatotoxicity under cypermethrin: Involvement of the crosstalk between Nrf2/KEAP1 and NF-κB/iκB-α pathways regulated by proteasome. Environmental Science & Technology. 2021;**55**(9):6171-6183

[29] Canning P, Sorrell FJ, Bullock AN. Structural basis of Keap1 interactions with Nrf2. Free Radical Biology and Medicine. 2015;**88**:101-107

[30] Dhanoa BS, Cogliati T, Satish AG, Bruford EA, Friedman JS. Update on the Kelch-like (KLHL) gene family. Human Genomics. 2013;**7**(1):1-7

[31] Jiang T, Harder B, De La Vega MR, Wong PK, Chapman E, Zhang DD. p62 links autophagy and Nrf2 signaling. Free Radical Biology and Medicine. 2015;**88**:199-204

[32] Kang KA, Hyun JW. Oxidative stress, Nrf2, and epigenetic modification contribute to anticancer drug resistance. Toxicological Research. 2017;**33**:1-5

[33] Kim ME, Ha TK, Yoon JH, Lee JS. Myricetin induces cell death

of human colon cancer cells via BAX/BCL2-dependent pathway. Anticancer Research. 2014;**34**(2):701-706

[34] Zimta AA, Cenariu D, Irimie A, Magdo L, Nabavi SM, Atanasov AG, et al. The role of Nrf2 activity in cancer development and progression. Cancers. 2019;**11**(11):1755

[35] Sharma P, Pattanshetty SM. A study on risk factors of cervical cancer among patients attending a tertiary care hospital: A case-control study. Clinical Epidemiology and Global Health. 2018;**6**(2):83-87

[36] Lee JM, Johnson JA. An important role of Nrf2-ARE pathway in the cellular defense mechanism. BMB Reports. 2004;**37**(2):139-143

[37] Leinonen HM, Kansanen E, Pölönen P, Heinäniemi M, Levonen AL. Role of the Keap1–Nrf2 pathway in cancer. Advances in Cancer Research. 2014;**122**:281-320

[38] Taguchi K, Yamamoto M. The KEAP1–NRF2 system in cancer. Frontiers in Oncology. 2017;7:85

[39] Chu XY, Li ZJ, Zheng ZW, Tao YL, Zou FX, Yang XF. KEAP1/NRF2 signaling pathway mutations in cervical cancer. European Review for Medical and Pharmacological Sciences. 2018;**22**(14):4458-4466

[40] Kansanen E, Kuosmanen SM, Leinonen H, Levonen AL. The Keap1-Nrf2 pathway: Mechanisms of activation and dysregulation in cancer. Redox Biology. 2013;**1**(1):45-49

[41] Silva-Islas CA, Maldonado PD. Canonical and non-canonical mechanisms of Nrf2 activation. Pharmacological Research. 2018;**134**:92-99

[42] Baird L, Swift S, Llères D, Dinkova-Kostova AT. Monitoring Keap1–Nrf2 interactions in single live cells. Biotechnology Advances. 2014;**32**(6):1133-1144

[43] Schmidlin CJ, Shakya A, Dodson M, Chapman E, Zhang DD. The intricacies of NRF2 regulation in cancer. In: Seminars in Cancer Biology. Vol. 76. Academic Press; 2021. pp. 110-119. DOI: 10.1016/j.semcancer.2021.05.016

[44] Culbreth M, Aschner M. GSK-3β, a double-edged sword in Nrf2 regulation: Implications for neurological dysfunction and disease. 2018;**F1000Research**:7

[45] Cominacini L, Mozzini C, Garbin U, Pasini A, Stranieri C, Solani E, et al. Endoplasmic reticulum stress and Nrf2 signaling in cardiovascular diseases. Free Radical Biology and Medicine. 2015;**88**:233-242

[46] He F, Antonucci L, Karin M. NRF2 as a regulator of cell metabolism and inflammation in cancer. Carcinogenesis. 2020;**41**(4):405-416

[47] Senkała S, Małecki JG, Vasylieva M, Łabuz A, Nosek K, Piwowarczyk K, et al. Hydrolysis of Schiff bases with phenyl-ethynyl-phenyl system: The importance for biological and physicochemical studies. Journal of Photochemistry and Photobiology B: Biology. 2020;**212**:112020

[48] Girisha M, Sagar BK, Yathirajan HS, Rathore RS, Kaur M, Jasinski JP, et al. Eight Schiff bases derived from various salicylaldehydes: Phenol–imine and keto–amine forms, conformational disorder, and supramolecular assembly in one and two dimensions. Acta Crystallographica Section C: Structural Chemistry. 2018;**74**(10):1094-1104

Chapter 9

Biological Potentials of Some Schiff Bases and Their Chelates: A Short Review

Ifeanyi Edozie Otuokere, Brendan Chidozie Asogwa, Felix Chigozie Nwadire, Onyinyechi Uloma Akoh, Chinedum Ifeanyi Nwankwo, Precious Onyinyechi Emole and Elias Emeka Elemike

Abstract

In the twenty-first century, the most significant task for scientists has been the development of novel and effective therapeutic agents. Many natural chemicals have been discovered and identified as having biological functions in the body that are beneficial to maintaining a healthy lifestyle. In addition, various classes of compounds are synthesized and assessed for their potential use in biological systems. Schiff bases have the capacity to react and form complexes with metals and proteins due to their superior basic characteristics; hence, the synthesis and pharmacological application of Schiff bases has recently become one of the most popular chemical and pharmaceutical science study subjects. As a result, this review discusses some of the novel Schiff bases and chelated complexes, as well as some proven pharmacological applications.

Keywords: Schiff bases, complexes, biological functions, structures, chelates

1. Introduction

Schiff bases are known with the structural formula $R_2C{=}NR_1$. They are prepared by the condensation of 1° NH_2 and C=O [1]. Hugo Schiff, a German Chemist who earned a Nobel Prize in 1864 [2], developed this group of chemicals. Schiff bases are carbonyl compounds where C=O is substituted with an azomethine or imine functionality in a reaction to generate a carbon to nitrogen double bond [3]. **Figure 1** shows Schiff's base structure.

Unsaturated -C=N- compounds that are easily coordinated to various metal ions are known as Schiff bases [4–6]. They have a flexible mode due to coordination (no set pattern) and a larger number of applications [7–10]. Due to the ease of their synthesis by condensing carbonyl compounds with amines, Schiff bases are frequently referred to as favored ligands [11]. They can also coordinate to a variety of metal ions with varying oxidation states. $R_2C{=}NR_1$ compounds are considered as a vital class of

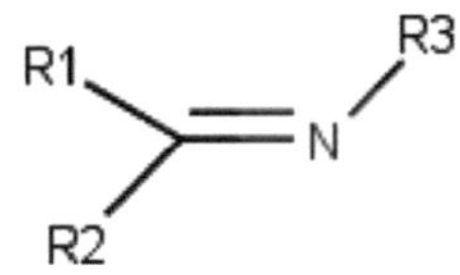

Figure 1.
Schiff's base structure.

organic compounds in the current development surge of medicinal compounds due to their vast applications in numerous disciplines [12]. This is because the azomethine groups attached to the nitrogen atom form a strong hydrogen bond [13] with different labile moieties and functional groups of various constituents of cell, interfering with or distorting vital cell metabolism and resulting in various desired effects [14]. The design of $R_2C{=}NR_1$ compounds was critical to the advancement of coordination (complex compound) chemistry, as well as inorganic biochemistry and many optical materials [15]. Metal complexes derived from chelating agents and $R_2C{=}NR_1$ compounds have been synthesized by a number of scientists [16–25].

1.1 Antimicrobials

Due to a number of factors, such as continuous emergence of contagious ailments and the explosion of various microbial infections that are multidrug resistant in nature, the search for an effective treatment material for various forms of contagious diseases has been and continues to be one among the most important challenges of the century. Despite the existence of a significant number of natural, synthetic antibiotics and chemotherapeutic substances for battling these diverse diseases, there has been an explosion of both old and new microorganisms that have become highly resistant to antibiotics developed in recent decades. This illustrates the critical need for classes of novel antimicrobial agents to be discovered and developed through diverse research and advancements [26].

The Knoevenagel condensate of keto esters (phenyl 2-(2-hydroxybenzylidene)-3-iminobutanoate) was used to prepare a Schiff base by reacting 2-[3-(3-hydroxynaphthalen-2-yl)-phenyl-3-imino-2-methylidenebutanoate]benzoic acid, and *o*-aminobenzoic acid in a molar ratio of 1:1. The Schiff base (Compound 1, **Figure 2**) was discovered to be biocidal against *Botrytis cinerea, Escherichia coli, Aspergillus niger*, *and Staphylococcus epidermidis*, implying that it has greater therapeutic (antimicrobial) efficacy than the relevant standards (ciprofloxacin and Co-trimoxazole) [3]. From benzaldehyde and sulfathiazole, a novel schiff base, 4-[(E)-(benzylideneamino)]-N-thiazol-2-yl-benzenesulfonamide, and its nickel chelate (Compounds 2 and 3 in **Figure 2**) were synthesized [27]. On a gram-negative strain of *Pseudomonas aeruginosa* and *E. coli*, and gram-positive strains of *Salmonella typhi and Staphylococcus aureus*, the new compounds' antibacterial activity was assessed *in vitro*. Both strains of bacteria demonstrated considerable antibacterial activity against the Schiff base. Overall, the metal complex demonstrated increased activity against bacterial strain, as it was discovered to be active against all bacterial strains tested [27]. Cu, Ni, Cd, Co, and Zn divalent chelates of the novel $R_2C{=}NR_1$ compound, 3-((4-phenylthiazol-2-ylimino) methyl)-2-hydroxybenzoic acid (Compounds 4 and 5 in **Figure 2**), have been reported [11]. The potential antibacterial activities of these new compounds were investigated. When compared to the new Schiff base ligand, the results revealed that the chelates were more antimicrobially active. The new compounds' activities involving DNA cleavage were also evaluated, and the plasmid DNA pBR 322 was strongly cleaved. The compounds' cytotoxicity was also assessed. The divalent Cu and Ni chelates were not cytotoxic, according to the results of the cytotoxicity tests [11].

The condensation of equivalent molar ratio of 1-(4-methylaniline)acetaldehyde oxime and 4-(diethylamino)-2-hydroxybenzaldehyde yielded a novel bidentate $R_2C{=}NR_1$ compound (E)-1-(4-((E)-(4-(*N*,*N*-diethylpropan-2-amine)-2-hydroxybenzylidene)amino)phenyl)ethanone oxime [28] (Compound 6, **Figure 2**). Using a 1:2 molar ratio of metal to ligand, a collection of chelates of the synthesized Schiff base with divalent Co, Ni, and Cu complexes (Compound 7, **Figure 2**) were also synthesized. The antimicrobial (*in vitro*) properties of the new $R_2C{=}NR_1$ compound

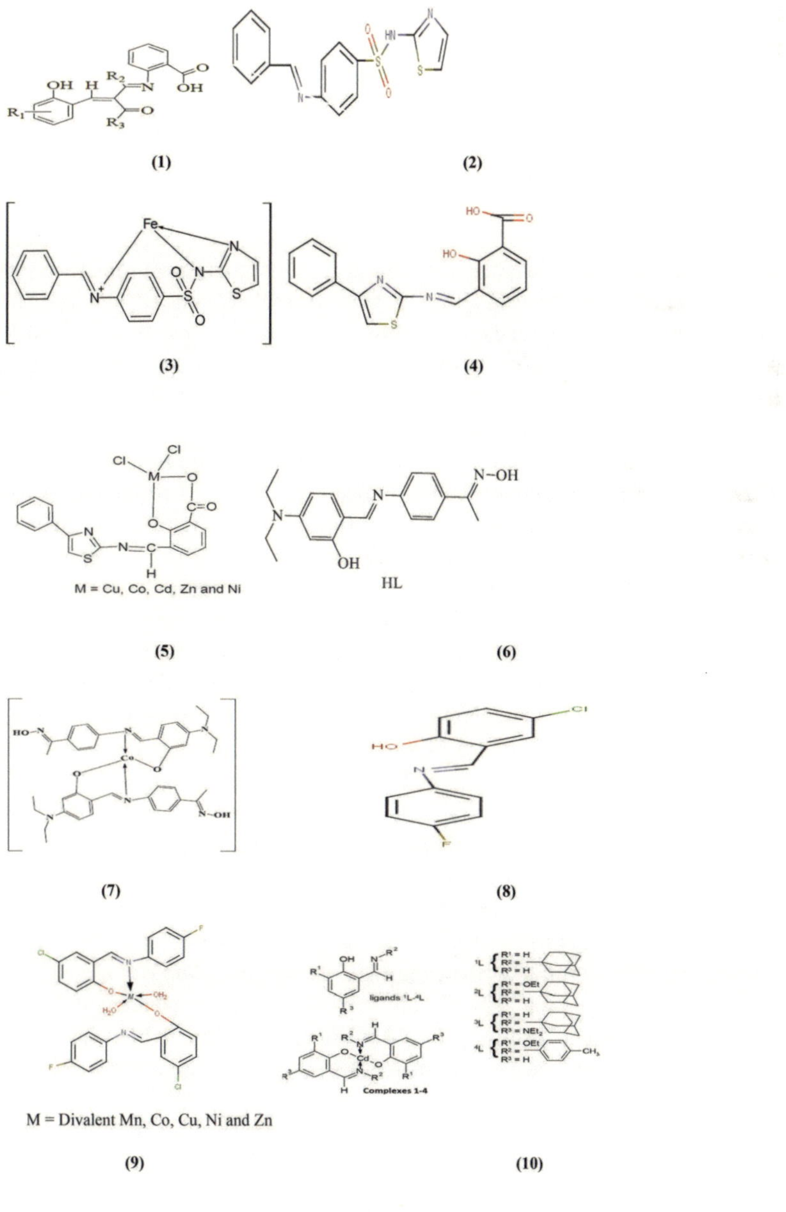

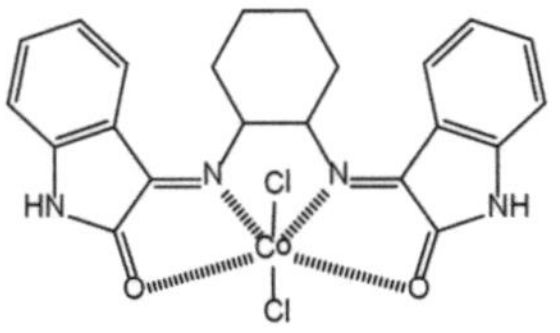

(11)

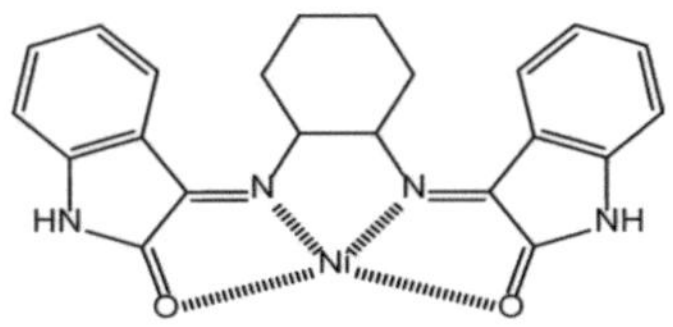

(12)

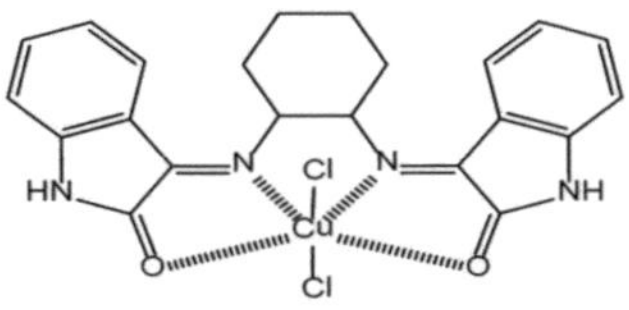

(13)

Figure 2.
Compounds showing antimicrobial activity.

and its chelates were investigated using the disk diffusion method, with dimethyl sulfoxide (DMSO) serving as a negative control. The study showed that the ligand, as well as its metal complexes, had equivalent inhibitory action against *Klebsiella* spp., *P. aeruginosa*, and *S. epidermidis* strains. The findings also revealed a significant inhibitory effect on harmful bacterial growth. The new compounds' inhibitory effects on the microbial fungus *Candida* were found to be greater at 10^{-4} M than at 10^{-3} M [28]. The divalent Mn, Co, Ni, Cu, and Zn complexes of the ligand (Compounds 8 and 9, **Figure 2**) were synthesized by the condensation of 5-chloro-2-hydroxybenzaldehyde with para-fluoro amino benzene at ambient temperature. The diffusion method using disks was applied to test the antibacterial potential of the novel $R_2C{=}NR_1$ compound and its coordination compounds against bacteria that are Gram-negative such as *P. aeruginosa* and *E. coli* as well as bacteria that are Gram-positive such as *S. typhi* and *B. subtilis*. The antibacterial assessment reports suggested that the chelates were more effective against bacteria than the uncoordinated Schiff base [29].

The condensation of tricyclo[3.3.1.13,7]decan-1-amine with three aromatic alkanals (2-hydroxybenzaldehyde, 3-ethoxy-4-hydroxybenzaldehyde, and 4-(diethylpropan-2-amine)2-hydroxybenzaldehyde) yielded a collection of novel $R_2C{=}NR_1$ compounds [30]. The divalent cadmium chelates of these $R_2C{=}NR_1$ compounds were also prepared [30]. The reaction of 4-methylaniline with 3-ethoxysalicyaldehyde yielded four Schiff bases [30] (Compound 10, **Figure 2**). The four Schiff base ligands that were synthesized had active antibacterial action against the tested bacterial strains, with *S. aureus* having the maximum activity. When compared to amikacin, the metal complexes had better antimicrobial action against *P. aeruginosa, Shigella flexneri, E. coli, S. typhi, B. subtilis,* and *S. aureus* [30]. Divalent Ni, Cu, and Co chelates of a novel $R_2C{=}NR_1$ complexing agent (Compounds 11–13, **Figure 2**) prepared from isatin and 1,2-cyclohexanediamine were tested for antibacterial potential against three different microbiological species such as *P. aeruginosa*, *E. coli,* and *S. aureus* [31]. The octahedral divalent Cu and Co chelates were shown to have significantly higher activity against microbial stains than the square planar Ni(II) complex and ligand [31].

DOI: http://dx.doi.org/10.5772/intechopen.114862

1.2 Antidepressant

Depression is a common psychiatric condition that affects about 21% of the world's population [32]. The physiological causes of depression are yet to be fully proven, despite repeated attempts to do so. However, some evidence suggested that depressive disorders could be caused by a lack of serotonin and noradrenaline. The majority of synthetic chemotherapeutic agents used to treat disorders like depression work by influencing the brain's biogenic amine system, resulting in the activation of a process capable of boosting biogenic amine concentration in the brain [33]. Oxidases of monoamine are an enzyme family that catalyzes the oxidation or inactivation of biogenic amines in the central nervous system (CNS) [34]. In addition to the wide range of antidepressant formulations and chemicals presently in the market, there are considerable proportions of patients who do not react to treatment or only have a partial response to drugs, necessitating the development of new antidepressants [34].

By using a standard approach, a novel Schiff base of indole-carrying azetidinone derivatives was synthesized with high purity and higher yields [35]. The compounds' derivatives (Compounds 14, **Figure 3**) showed antidepressant activity that was comparable to that of fluoxetine, a well-known antidepressant. The ortho-position of the phenyl ring system was replaced with nitrogen and chlorine, resulting in active compounds with immobility times of 66.82 and 65.61%, respectively. The results of the evaluation of molecular docking studies were in agreement with the pharmacological studies with a docking score of −2.8474 kcal/mol [35].

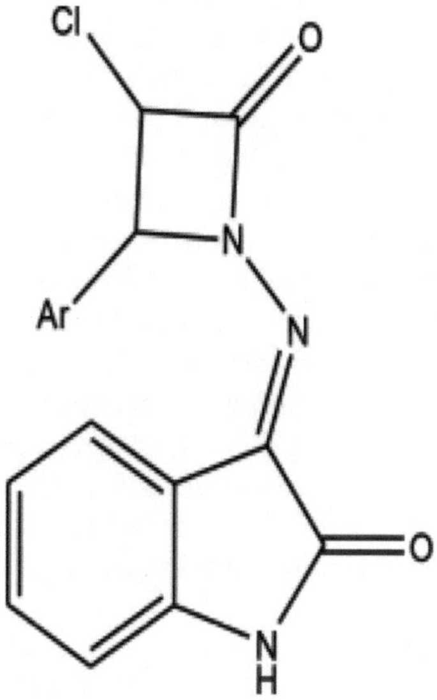

Ar	Ar
C_6H_5-	4-CH_3-C_6H_5-
4-Cl-C_6H_5-	2-OCH_3-C_6H_5-
4-C_6H_5-CH_2O-C_6H_5-	4-OCH_3-C_6H_5-
2-NO_2-C_6H_5-	2-Cl-C_6H_5-
4-OH-C_6H_5-	3-Br-C_6H_5-
4-OH,3-OCH_3-C_6H_5-	2,4-$(OCH_3)_2$-C_6H_5-
4-$(CH_3)_2$-N-C_6H_5-	4-NO_2-C_6H_5-
3-NO_2-C_6H_5-	2,3-$(OCH_3)_2$-C_6H_5-
3,4,5-$(OCH_3)_3$-C_6H_5-	3-Cl-C_6H_5-
2,4-$(Cl)_2$-C_6H_5-	C_4H_3O- (2-furyl)

(14)

Figure 3.
Compounds showing antidepressant properties.

1.3 Antitubercular

Mycobacterium tuberculosis is a bacterium that causes tuberculosis infection [36]. The prevalence or dominance of tuberculosis, which has been on the rise in recent years as tuberculosis is resistant to a variety of drugs [37, 38], has created the need for a broad search for more potent therapeutic agents with lesser negative effects that could help control its spread. Tuberculosis resisting therapy and management have proven to be a difficult task in recent years, as second-line drugs have generally become ineffective against them [39]. *M. tuberculosis* is a bacterium that is distinguished by its thick and waxy cell wall.

The condensation cyclorization of 2-amino-5-phenyl-5H-thiazolo[4,3-b]-l,3,4-thiadiazole with various aromatic alkanals resulted in a collection of $R_2C=NR_1$ complexing agents [40]. The antitubercular activity (*in vitro*) of these $R_2C=NR_1$ complexing agents (Compound 15, **Figure 4**) was assessed using the method of Microplate Alamar Blue Assay. Streptomycin and pyrazine-2-carboxamide antitubercular drugs at varied concentrations ranging from 0.1 to 100.0 g/mL were used as standards. Two of the newly synthesized compounds (labeled 4f and 4i) had a larger molecular weight than the others. When compared to pyrazinamide and streptomycin (standards), two of the novel synthesized Schiff bases showed a comparatively higher antitubercular activity, while four other Schiff bases exhibited action that was exactly similar to that of pyrazine-2-carboxamide, but showed lower effectiveness compared to streptomycin [40]. When compared to traditional medications, molecular docking result implied that the new Schiff bases had a decent docking score. The antitubercular activity of 4-amino-5-(4-fluoro-1,1′-oxydibenzene)-4*H*-1,2,4-triazole-3-thiol and its synthesized $R_2C=NR_1$ complexing agents were tested against H37Rv and multidrug-resistant *Mycobacterium* tuberculosis (MDR MTB) strains at concentrations ranging from 0.2 to 32 g/mL (Compounds 16–19, **Figure 4**). The antitubercular activity obtained revealed that the parent compound had the most promising antitubercular activity, with inhibition concentrations of 50 μg/ml at a concentration of 5.5 and 11.0 g/mL. The remaining compounds (Compounds 17, 18, and 19) were exclusively active against H37Rv, indicating that they did not have good antitubercular efficacy against MDR strains up to 32 g/mL concentrations. Compounds 17 and 19 had similar antitubercular effects against MTB strains H37Rv at doses of 2 g/mL [40].

By reacting ethyl-2-amino-4,5,6,7-tetrahydrobenzo-1-carboxylato-1*H*-1λ^4-thien-1-yl and salicylaldehyde derivatives, Bootwala and coworkers [41] developed two active Schiff bases. Streptomycin, pyrazinamide, and ciprofloxacin were used as standards to test these drugs for antitubercular efficacy against *Mycobacterium tuberculosis* [41]. According to the findings, the new Schiff base compounds were the most active antitubercular compounds against *Mycobacterium* TB, whereas the metal complexes show only minor action against the bacteria, with minimum inhibitory doses ranging from 12.5 to 50.0 g/ml.

1.4 Anthelmintic

The anthelmintic activity of some substituted 1,3,4-thiadiazole Schiff bases synthesized from benzimidazole (Compound 22, **Figure 5**) was investigated [42]. The new compounds had a mean paralyzing time (min) of 9.10–18.23 and 13.11–22.31 min against *Perionyx excavatus* and *Pheretima posthuma*, respectively, indicating good action when compared to albendazole, which had a mean paralyzing time of 9.70 and 12.20 min. The evaluated drugs' mean death times (min) against *P. excavatus* and *P.*

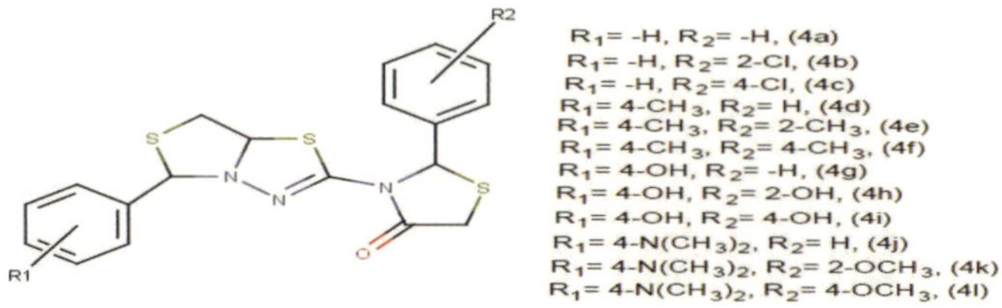

(15)

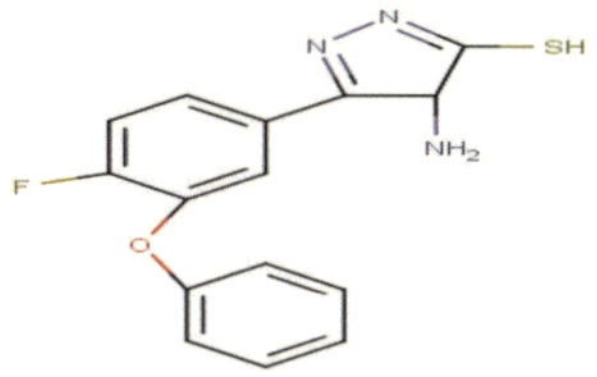

(16)

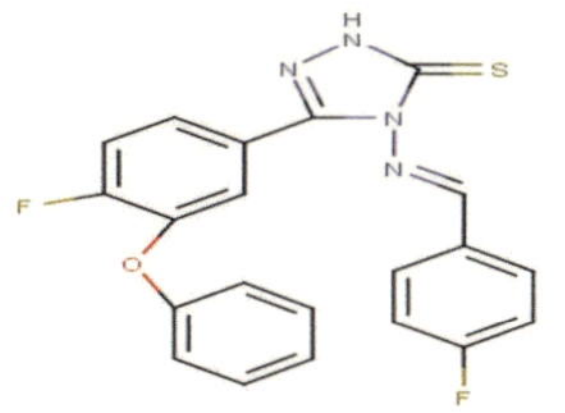

(17)

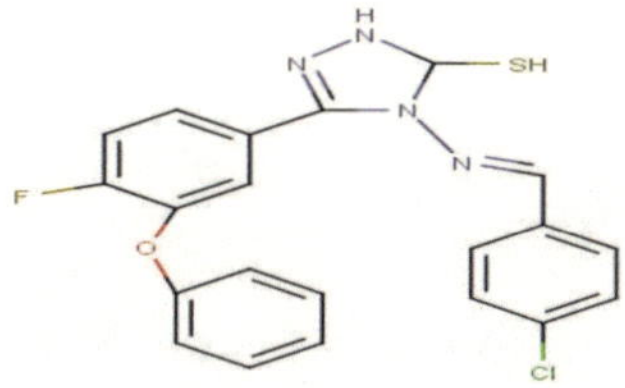

(18)

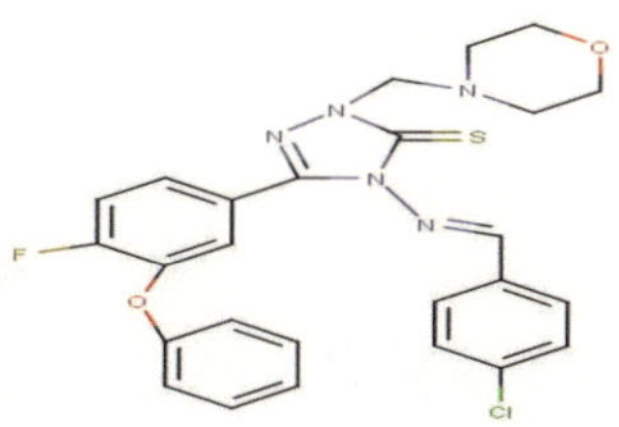

(19)

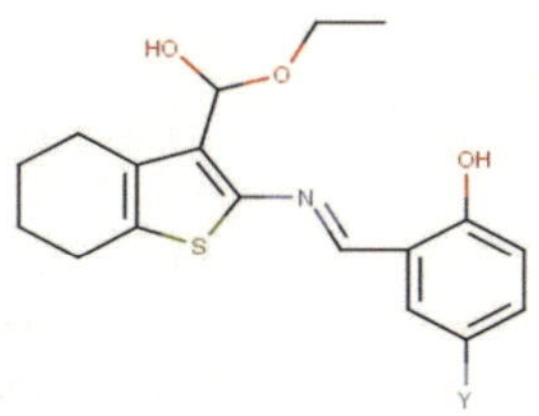

Where Y = Cl, Br

(20)

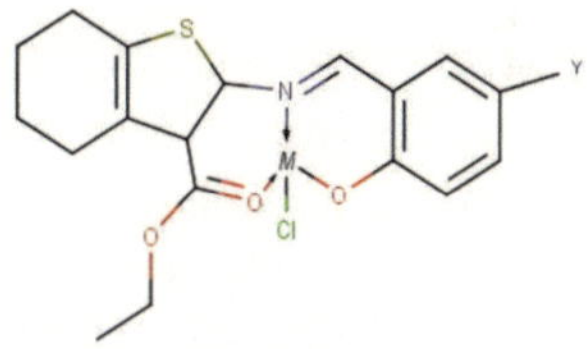

Where M = Ni, Cu, Zn, Y = Cl, Br

(21)

Figure 4.
Compounds with antitubercular activities.

posthuma varied from 13.22–28.36 to 19.00–28.32 min, which is equivalent to albendazole's 14.80 and 20.70 min mean death time. The compound with the $4NO_2$ substitution was discovered to be more effective than albendazole in killing nematodes, taking an average of 13.22 and 19.00 minutes to kill *P. excavatus* and *P. posthuma*, respectively. The anthelmintic potential of the derivatives containing bromine (Br) and OCH_3 is likewise extremely similar to that of albendazole. The results showed that derivatives having electron-drawing moieties at the ring's *o*-position had a significant increase in activity, as did derivatives with hydroxyl substituents in the ring's *p*-position and CH_3O substituents in the ring's *o*-position. As a result of the aforesaid findings, it was deduced that the

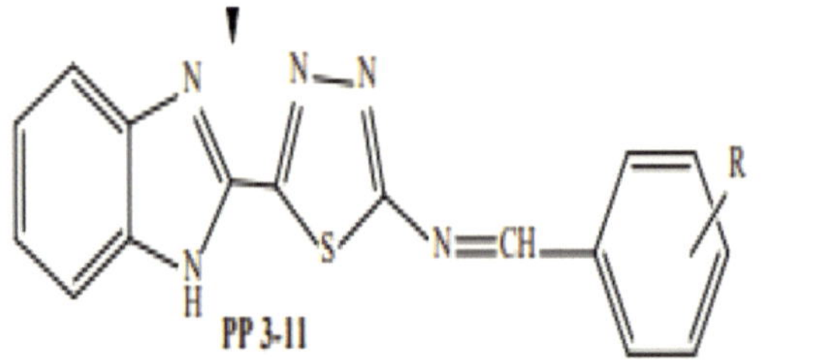

R= H, 4-NO_2, 2-OH, 2-Br, 4-OH, 4-Cl, 4-CH_3, 4-OCH_3, 4-OH and 2-OCH_3

(22)

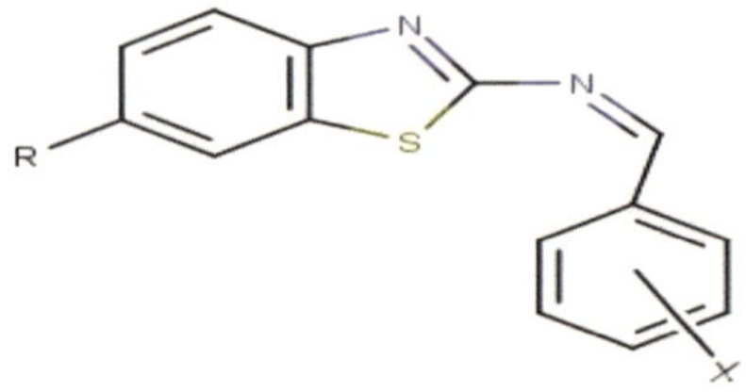

R = F and NO_2

X = Cl , OCH_3 , $N(CH_3)_2$, Vanilline R = F and NO_2

(23)

Figure 5.
Compounds that showed anthelmintic activity.

electron-withdrawing group performed a crucial role in the anthelmintic potential of the compounds [42]. Using conventional methods, a series of five fluoro- and nitro- (F- and NO_2-)substituted ethenylbenzene-2 amino-1,3-benzothiazole and its derivatives (Compound 23, **Figure 5**) were prepared from F^- and NO_2-substituted 2-aminobenzothiazole with some aromatic alkanals [43]. Compound 23 was tested for antibacterial and anthelmintic properties *in vitro* using albendazole as a control. Compound 23 (a and f) had a paralysis time of 12 and 10 minutes, and a death time of 14 and 13 minutes, respectively, compared to 22 and 24 minutes for ordinary albendazole [43].

1.5 Antioxidant and antitumor

The azomethine compound, 2-(2-oxoacenaphthylen-1(2*H*)-ylidene)-*N*-(prop-2-en-1-yl)hydrazinecarbothioamide, and its divalent Zn, Co, and Ni (Compounds 24–26, **Figure 6**) derivatives were prepared and evaluated for antioxidant and antitumor properties [44]. The results implied that the complexing agent and divalent Zn chelate have higher antioxidant potential than other chelates, with values of 88.5 and 88.6%, respectively, when compared to a standard ascorbic acid. The compound's antitumor potential against the cell line of hepatocellular carcinoma (HepG2) revealed that both the Zn(II) complex and Schiff base ligand showed considerable

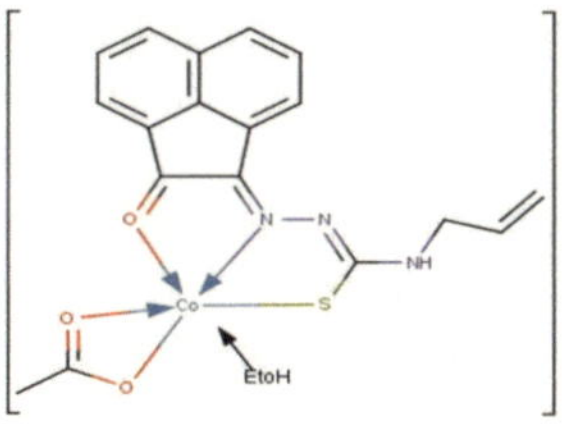

(24)

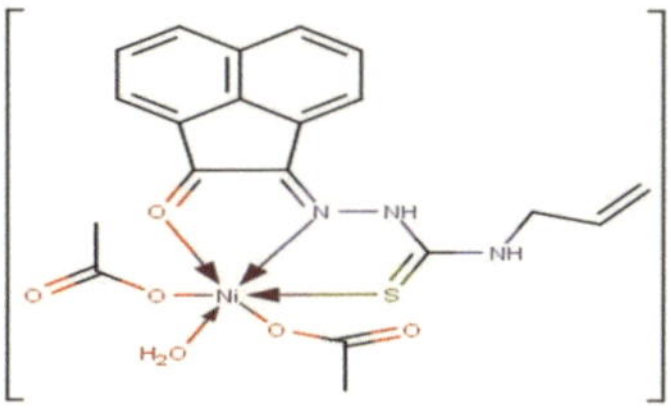

(25)

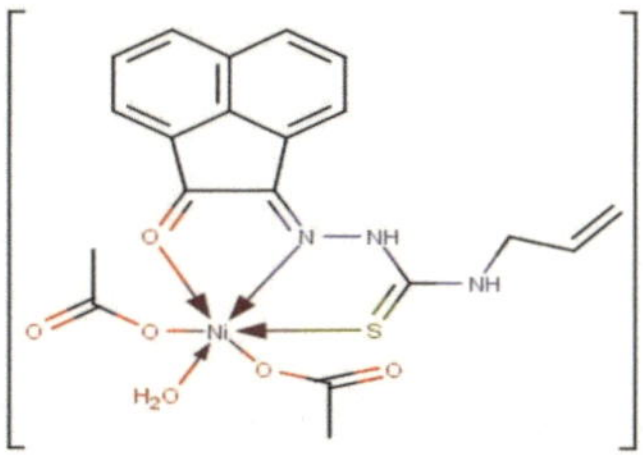

(26)

Figure 6.
Compounds having antioxidant and antitumor properties.

activity, with 6.39 ± 0.18 and 6.45 ± 0.25 μM IC_{50} (half-maximal inhibitory concentration) values, respectively [44].

1.6 Anticonvulsant

Epilepsy is a neurological condition characterized by abnormal electrical impulses in the CNS. It causes seizures that occur spontaneously and repeatedly, with loss or gain of consciousness [45]. Six Schiff bases (Compound 27, **Figure** 7) were synthesized from 1*H*-indole-2,3-dione and tricyclo[3.3.1.13,7]decane-1-carbohydrazide and were investigated for efficacy of anticonvulsants in a 6,7,8,9-tetrahydro-5*H*-tetrazolo[1,5-*a*]azepine-induced seizure paradigm using 5-benzyl-5-methylpyrimidine-2,4,6(1*H*,3*H*,5*H*)-trione as a positive reference standard [46]. Pentylenetetrazole (PTZ)-induced seizures were significantly reduced by these synthesized Schiff bases. The delay of recurrent convulsions was

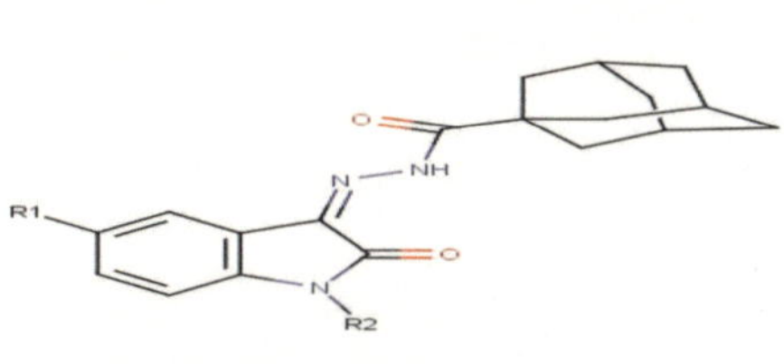

27

R1 = H, Br, Cl R2 = H, CH_2OH, CH_2Ph

Figure 7.
Compounds having anticonvulsant properties of the SARS CoV 2 (severe acute respiratory syndrome coronavirus 2) inhibitor.

increased in all synthesized compounds, and the length of time that epilepsy lasted was reduced, as well as neurological symptoms and seizure intensity [46].

The SARS-CoV-2 virus, which produced the unique coronavirus epidemic known as Coronavirus 2019 (COVID-19), quickly expanded throughout the world, with Wuhan, China, serving as its epicenter by the end of 2019 [47–50]. The virus, called coronavirus, has a diameter of 60–140 nm and resembles spike glycoproteins in the form of a crown when viewed under an electron microscope [50–53]. On March 11, 2020, the World Health Organization (WHO) declared the COVID-19 epidemic to be a pandemic due to its high rate of infection, intensity, and rate of spread [54]. Currently, COVID-19 is a problem in over 225 countries, and the number of infections is still growing quickly. This coronavirus epidemic has not only put people's health at danger but has also had a significant negative impact on the global economy [55, 56].

Vaccines, however, are being researched on children and have been found to be safe for adults [57]. Once more, virus mutations or alterations could render vaccinations useless [58]. Drugs to treat those infected with SARS-CoV-2 are still required, even though vaccination rates have decreased. Nevertheless, a number of local and international research organizations are still working to find a COVID-19 therapy medicine or vaccine that works while taking into account the risk factors related to viral infection. New chemotherapeutics have caught the interest of medicinal chemists in recent years. Isatin's schiff base ligands have been shown to exhibit antiviral properties against the SARS, vaccinia, rhino, and Moloney leukemia viruses [59]. Considerable antiviral activity was demonstrated by Schiff bases made from 5-acyl-1,2,4-triazines containing oximes, hydrazones, semicarbazones, and thiosemicarbazones [60]. Some

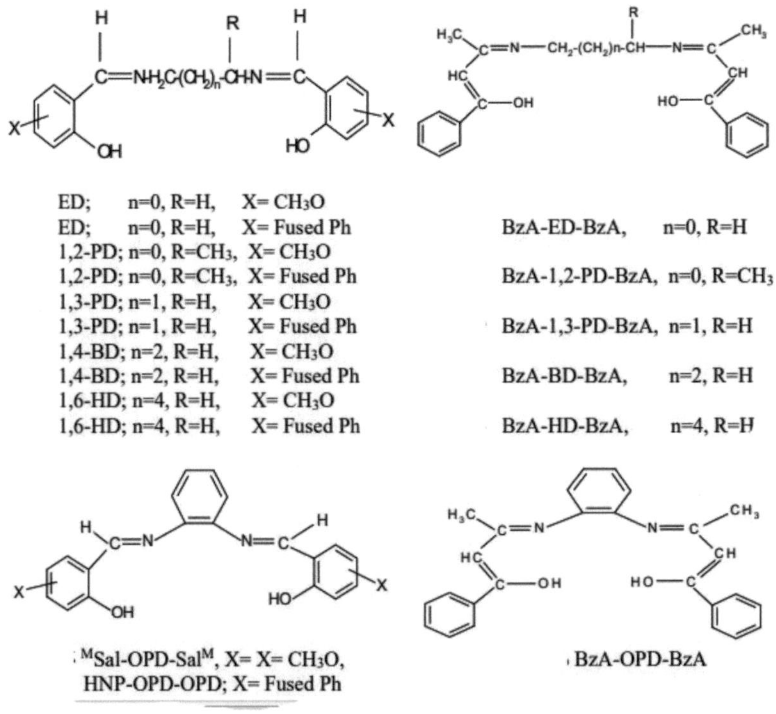

Figure 8.
Proposed SARS CoV 2 inhibitors.

1,6-hexanediamine tetra-dentate symmetrical bis-Schiff bases were predicted *in silico* to be potential coronavirus (SARS-CoV-2) inhibitors [61]. Some previously synthesized Schiff base ligands have been optimized using density functional theory, and their physicochemical characteristics have been studied [62–64]. In order to comprehend the interactions and binding affinities between ligands and receptor proteins, molecular docking has been used. To determine their biological position, including absorption, metabolism, and carcinogenicity, the ADMET (Absorption, Distribution, Metabolism, Excretion and Toxicity) analysis was performed. It has been reported that some schiff bases have been evaluated *in silico* for their potency against the SARS-CoV-2 main protease (Mpro), PASS (prediction of activity spectra for substances), and ADMET studies (Compounds 28, **Figure 8**) [65].

2. Conclusion

Schiff bases are a type of chelating agent that can be found in a variety of drugs. There has been a lot of research on Schiff bases. This study emphasized the importance of $R_2C{=}NR_1$ compounds in the advancement of novel compounds with medicinal applications. Regardless that study in this field is still in its infantile phase, there have recently been surges in reports demonstrating the impact of $R_2C{=}NR_1$ compounds on therapeutically important microorganisms. For decades, because of this bioactive core, researchers have been interested in obtaining several $R_2C{=}NR_1$ compounds and chelates of biological significance. The goal of this study was to look at all of the biological information available.

Author details

Ifeanyi Edozie Otuokere[1], Brendan Chidozie Asogwa[1], Felix Chigozie Nwadire[1], Onyinyechi Uloma Akoh[1], Chinedum Ifeanyi Nwankwo[2], Precious Onyinyechi Emole[3] and Elias Emeka Elemike[4*]

1 Department of Chemistry, Michael Okpara University of Agriculture, Umudike, Nigeria

2 Department of Biochemistry, Michael Okpara University of Agriculture, Umudike, Nigeria

3 Department of Chemistry, Abia State University, Uturu, Nigeria

4 Department of Chemistry, Federal University of Petroleum Resources Effurun, Effurun, Nigeria

*Address all correspondence to: chemphilips@yahoo.com

References

[1] Zhang S, Jiang Y, Chen M. Synthesis and antibacterial activity of N-5 chlorosalicylidene taurine Schiff base and its Cu and Ni complexes. Chemical Abstract. 2005;**22**:59-61

[2] Schiff H. Mittheilungen aus dem universita¨tslaboratorium in Pisa: Eine neue reihe organischer basen. Justus Liebigs Annalen der Chemie. 1864;**131**(1):118-119

[3] Kalaivani S, Priya NP, Arunachalam S. Schiff bases: Facile synthesis, spectral characterization and biocidal studies. IJABPT. 2012;**3**:219-223

[4] Petrus M, Bouwer R, Lafont U, Athanasopoulos S, Greenham N, Dingemans T. Small-molecule azomethines: Organic photovoltaics via Schiff base condensation chemistry. Journal of Materials Chemistry. 2014;**2**:9474-9477

[5] Ilhan S, Baykara H, Seyitoglu M, Levent A, Özdemir S, Dündar A, et al. Preparation, spectral studies, theoretical, electrochemical and antibacterial investigation of a new Schiff base and some of its metal complexes. Journal of Molecular Structure. 2014;**1075**:32-42

[6] Gaber M, El-Wakiel NA, El-Ghamry H, Fathalla SK. Synthesis, spectroscopic characterization, DNA interaction and biological activities of the Mn (II), Co (II), Ni (II) and Cu (II) complexes with [(1H-1,2,4-triazole-3-ylimino) methyl] naphthalene-2-ol. Journal of Molecular Structure. 2014;**1076**:251-261

[7] Jia L, Xu I, Zhao X, Shen S, Zhou T, Xu Z, et al. Synthesis, characterization, and antitumor activity of three ternary dinuclear copper (II) complexes with a reduced Schiff base ligand and diimine coligands in vitro and in vivo. Journal of Inorganic Biochemistry. 2016;**159**:107-119

[8] Bensaber SM, Allafe H, Ermeli NB, Mohamed SB, Zetrini AA, Alsabri SG, et al. Chemical synthesis, molecular modelling, and evaluation of anticancer activity of some pyrazol-3-one Schiff base derivatives. Medicinal Chemistry Research. 2014;**23**:5120-5134

[9] Kumar S, Ganguly R, Veerasamy KS, De Clercq E. Synthesis, antiviral activity and cytotoxicity evaluation of Schiff bases of some 2-phenyl quinazoline-4(3)H-ones. European Journal of Medicinal Chemistry. 2010;**45**(11):5474-5479

[10] Shokohi-Pour Z, Chiniforoshan H, Momtazi-Borojeni AA, Notash B. A novel Schiff base derived from the gabapentin drug and copper (II) complex: Synthesis, characterization, interaction with DNA/protein and cytotoxic activity. Journal of Photochemistry and Photobiology. 2016;**162**:34-44

[11] Kalshetty M, Karabasannavar BS, Allolli P, Shaikh IN. Synthesis and antimicrobial activity of some metal complexes. Indian Journal Of Pharmaceutical Education And Research. 2017;**51**:3

[12] Kajal A, Bala S, Kamboj S, Sharma N, Saini V, Bases S. A versatile pharmacophore. Journal of Catalysts. 2013;**2013**:893512

[13] Venugopala KN, Jayashree BS. Synthesis of carboxamides of 2`-amino-4`-(6-bromo-3-coumarinyl) thiazole as analgesic and antiinflammatory agents. Indian Journal of Heterocyclic Chemistry. 2003;**12**(4):307-310

[14] Vashi V, Naik H. Synthesis of novel Schiff base and azetidinone derivatives and their antibacterial activity. European Journal of Chemistry. 2004;**1**:272-276

[15] Tisato F, Refosco F, Bandoli G. Structural survey of technetium complexes. Coordination Chemistry Reviews. 1994;**135-136**:325-397

[16] Otuokere IE, Anyanwu JC, Igwe Synthesis KK. Characterization and antibacterial studies of 4-[(E)-benzylideneamino]-Nthiazoyl-2-yl-benzenesulfonamide and its Fe(III) complex, Nigerian research. Journal of Chemical Sciences. 2021;**9**(1):173-185

[17] Eberendu KO, Otuokere IE. Complexation behavior of 1-Phenyl-2,3-dimethyl-4-(benzylamino)pyrazol-5-one Schiff Base ligand with manganese ion. Research Journal of Science and Technology. 2016;**8**(4):199-203

[18] Onyenze U, Otuokere IE, Igwe JC. Co (II) and Fe (II) mixed ligand complexes of Pefloxacin and ascorbic acid: Synthesis, characterization and antibacterial studies. Research Journal of Science and Technology. 2016;**8**(4):215-220

[19] Otuokere IE, Amadi KC. Synthesis and Characterization of Glimepiride Yttrium Complex. International Journal of Pharmaceutical Sciences and Drug Research. 2017;**1**(2):10-14

[20] Otuokere IE, Onyenze U, Obike AI, Ahamefula AA, Okafor GU. Synthesis, characterization and antibacterial studies of Co (II) and Fe(II) mixed ligand complexes of ciprofloxacin and ascorbic acid. The Journal of Environmental Studies and Sciences. 2017;**2**(1):12-20

[21] Otuokere IE, Robert UF. Synthesis, characterization and antibacterial studies of (3,3-dimethyl)-7-oxo-6-(2-phenylacetamido)-4-thia-1-azabicyclo[3,2,0]heptanes-2-carboxylic acid-Cr(III) complex. The Journal of Nepal Chemical Society. 2020;**41**(1):1-7

[22] Otuokere IE, Anyanwu JC, Igwe KK. Synthesis, characterization and antibacterial studies of 4-{[(E)-Phenylmethylidene] amino}-N-(1,3-thiazol-2-yl)benzenesulfonamide and its Mn(II) complex. ChemSearch Journal. 2020;**11**(1):44-512

[23] Amadi OK, Otuokere IE, Bartholomew CF. Synthesis, Characterization, in vivo Antimalarial Studies and Geometry Optimization of Lumefantrine/Artemether Mixed Ligand Complexes. Research Journal of Pharmaceutical Dosage Forms and Technology. 2015;**7**(1):59-68

[24] Amadi OK, Otuokere IE, Chinedum NC. Synthesis and antibacterial studies of Fe(II), Co(II), Ni(II) and Cu(II) complexes of Lumefantrine. International Journal of Chemical, Gas and Material Science. 2017;**1**(1):6-13

[25] Otuokere IE, Amadi KC. Synthesis and characterization of glimepiride yttrium complex. International Journal of Medical and Pharmaceutical Research. 2017;**1**:10-13

[26] Abdulkarem AA. Synthesis and antibacterial studies of metal complexes of Cu(II), Ni(II) and Co(II) with Tetradentate ligand. Journal of Biophysical Chemistry. 2017;**8**:13-21

[27] Otuokere IE, Anyanwu JC, K. K. Igwe Synthesis, spectra and antibacterial studies of 4-{[(*E*)-phenylmethyl- idene]amino}-*N*-(1,3-thiazol-2-yl)benzenesulfonamide Schiff Base ligand and its Ni(II) complex. Communications in Physical Sciences. 2021;**7**(2):115-125

[28] Al-Zaidi BH. New complexes of chelating Schiff base: Synthesis, spectral investigation, antimicrobial, and thermal behavior studies. Journal of Applied Pharmaceutical Sciences. 2019;**9**(04):045-057

[29] Ommenya FK, Nyawade EA, Andala DM, Kinyua J. Synthesis, characterization and antibacterial activity of Schiff Base, 4-Chloro-2-{(E)-[(4-Fluorophenyl)imino]methyl}phenol metal (II) complexes. Hindawi Journal of Chemistry. 2020;**8**:1745236

[30] Irfan ARM, Shaheen MA, Saleem M, Tahir MN, Munawar KS, Ahmad S, et al. Synthesis of new cadmium(II) complexes of Schiff bases as alkaline phosphatase inhibitors and their antimicrobial activity. Arabian Journal of Chemistry. 2021;**14**:103308

[31] Rajendrakumar KD, Menghani SS, Khedekar PB. Synthesis, characterization, antidepressant and docking studies of some novel indole bearing azetidinone derivatives as antidepressants. Indian Journal Of Pharmaceutical Education And Research. 2018;**52**(1):110-121

[32] Tipton KF, O'Sullivan BS, Davey GP, Healy J. Monoamine oxidases: Certainties and uncertainties. Current Medicinal Chemistry. 2004;**11**(15):1965-1982

[33] Edmondson DE, Mattevi A, Binda C, Li M, Hubálek F. Structure and mechanism of monoamine oxidase. Current Medicinal Chemistry. 2004;**11**(15):1983-1993

[34] Tran PV, Bymaster FP, McNamara RK, Potter WZ. Dual monoamine modulation for improved treatment of major depressive disorder. Journal of Clinical Psychopharmacology. 2003;**23**(1):78-86

[35] Katharigatta NV, Kandeel M, Pillay M, Deb PK, Abdallah HH, Mahomoodally MF, et al. Anti-tubercular properties of 4-Amino-5-(4-Fluoro-3-Phenoxyphenyl)-4H-1,2,4-Triazole-3-thiol and its Schiff Bases: Computational input and molecular dynamics. Antibiotics. 2020;**9**:559

[36] Marcos AE. The global situation of MDR-TB. Tuberculosis. 2003;**83**:44-51

[37] Caminero JA, Sotgiu G, Zumla A, Migliori GB. Best drug treatment for multidrug-resistant and extensively drug-resistant tuberculosis. The Lancet Infectious Diseases. 2010;**10**:621-629

[38] Hu Y, Xu L, He YL, Pang Y, Lu N, Liu J, et al. Prevalence and molecular characterization of second-line drugs resistance among multidrug-resistant mycobacterium tuberculosis isolates in southwest of China. Bio Med Research International. 2017;**17**:4563826

[39] Chang K, Lu W, Wang J, Zhang K, Jia S, Li F, et al. Rapid and effective diagnosis of tuberculosis and rifampicin resistance with Xpert MTB/RIF assay: A meta-analysis. The Journal of Infection. 2012;**64**(6):580-588

[40] Himaja M, Karigar AA, Malipeddi VR, Sikarwar MS. Synthesis and Antitubercular activity of some novel Thiazolidinone derivatives. Tropical Journal of Pharmaceutical Research. 2012;**11**(4):611-620

[41] Sakina B, More G, Shenoy S, Mascarenhas J, Aruna K. Synthesis, characterization and In vitro Antitubercular and antimicrobial activities of new Aminothiophene Schiff Bases and their Co(II), Ni(II), Cu(II) and Zn(II) metal complexes. Oriental Journal of Chemistry. 2018;**34**(2):800-812

[42] Saravanan K, Priyabrata P. Synthesis, characterization and In vitro evaluation

for antimicrobial and anthelmintic activity of novel Benzimidazole substituted 1,3,4-Thiadiazole Schiff's Bases. FABAD Journal of Pharmaceutical Sciences. 2021;**46**(3):1-20

[43] Balaji PNN, Chanukya AK, Mohana C, Bindu KH, Priyanka S. *In vitro* anthelmintic and antimicrobial activity of synthesized fluoro & nitro substituted N-[(Z)-phenylmethylidene]-1,3-benzothiazol-2- amine and its derivatives. Der Chemica Sinica. 2014;**5**(1):110-114

[44] Khlood S, Melha A, Al-Hazmi GA, Althagafi I, Alharbi A, Keshk AA, et al. Spectral, molecular Modeling, and biological activity studies on new Schiff's base of Acenaphthaquinone transition metal complexes. Hindawi Bioinorganic Chemistry and Applications. 2021;**17**:6674394

[45] Saravanan G, Alagarsamy V, Dineshkumar P. Anticonvulsant activity of novel 1-(morpholinomethyl)-3-substituted isatin derivatives. Bulletin of Faculty of Pharmacy, Cairo University. 2014;**52**(1):115-124

[46] Osman HM, Tilal E, Bashir AY, Esraa E, Aimun AE, Eyman ME, et al. Schiff Bases of Isatin and Adamantane-1-Carbohydrazide:Synthesis, characterization and anticonvulsant activity. Hindawi Journal of Chemistry. 2021;**2021**(1):665915

[47] Lai CC, Shih TP, Ko WC, Tang HJ, Hsueh PR. Severe acute respiratory syndrome coronavirus 2 (SARS-CoV-2) and coronavirus disease-2019 (COVID-19): The epidemic and the challenges. International Journal of Antimicrobial Agents. 2020;**55**(3):105924. DOI: 10.1016/j.ijantimicag.2020.105924

[48] Chen Y, Liu Q, Guo D. Emerging coronaviruses: Genome structure, replication, and pathogenesis. Journal of Medical Virology. 2020;**92**(4):418-423. DOI: 10.1002/jmv. 25681

[49] Singh AK, Singh A, Shaikh A, Singh R, Misra A. Chloroquine and hydroxychloroquine in the treatment of COVID-19 with or without diabetes: A systematic search and a narrative review with a special reference to India and other developing countries. Diabetes & Metabolic Syndrome: Clinical Research & Reviews. 2020;**14**(3):241-246. DOI: 10.1016/j.dsx.2020.03.011

[50] Ghosh R, Chakraborty A, Biswas A, Chowdhuri S. Evaluation of green tea polyphenols as novel coronavirus (SARS CoV-2) main protease (Mpro) inhibitors–an in silico docking and molecular dynamics simulation study. Journal of Biomolecular Structure and Dynamics. 2021;**39**(12):4362-4374. DOI: 10.1080/07391102.2020.1779818

[51] Boopathi S, Poma AB, Kolandaivel P. Novel 2019 coronavirus structure, mechanism of action, antiviral drug promises and rule out against its treatment. Journal of Biomolecular Structure and Dynamics. 2021;**39**(9):3409-3418. DOI: 10.1080/07391102.2020.1758788

[52] Grum-Tokars V, Ratia K, Begaye A, Baker SC, Mesecar AD. Evaluating the 3C-like protease activity of SARS-coronavirus: Recommendations for standardized assays for drug discovery. Virus Research. 2008;**133**(1):63-73. DOI: 10.1016/j.virusres.2007.02.015

[53] Harcourt BH, Jukneliene D, Kanjanahaluethai A, Bechill J, Severson KM, Smith CM, et al. Identification of severe acute respiratory syndrome coronavirus replicase products and characterization of papain-like protease activity. Journal of Virology. 2004;**78**(24):13600-13612. DOI: 10.1128/JVI.78.24.13600-13612.2004

[54] Osman EEA, Toogood PL, Neamati N. COVID-19: Living through another pandemic. ACS Infectious Diseases. 2020;**6**(7):1548-1552. DOI: 10.1021/acsinfecdis. 0c00224

[55] Yan L, Velikanov M, Flook P, Zheng W, Szalma S, Kahn S. Assessment of putative protein targets derived from the SARS genome. FEBS Letters. 2003;**554**(3):257-263. DOI: 10.1016/ S0014- 5793(03)01115-3

[56] Yao X, Ye F, Zhang M, Cui C, Huang B, Niu P, et al. *In vitro* antiviral activity and projection of optimized dosing design of hydroxychloroquine for the treatment of severe acute respiratory syndrome coronavirus 2 (SARS-CoV-2). Clinical Infectious Diseases. 2020;**71**(15):732-739. DOI: 10.1093/cid/ciaa237

[57] INA F, SSBE. A. COVID vaccines and kids: Five questions as trials begin. Nature. 2021;**592**(7856):670-671. DOI: 10.1038/d41586-021-01061-4

[58] Kupferschmidt K. New Coronavirus Variants Could Cause More Reinfections, Require Updated Vaccines. USA: American Association for the Advancement of Science; 2021. DOI: 10.1126/science.abg6028

[59] Chen LR, Wang YC, Lin YW, Chou SY, Chen SF, Liu LT, et al. Synthesis and evaluation of isatin derivatives as effective SARS coronavirus 3CL protease inhibitors. Bioorganic & Medicinal Chemistry Letters. 2005;**15**(12):3058-3062. DOI: 10.1016/j.bmcl.2005.04.027

[60] Mojzych M, Sebela M. Synthesis and biological activity evaluation of Schiff bases of 5-acyl-1,2,4-triazine. Journal of the Chemical Society of Pakistan. 2015;**37**(2):300-305

[61] Uddin MN, Amin MS, Rahman MS, Khandaker S, Shumi W, Rahman MA, et al. Titanium (IV) complexes of some tetra-dentate symmetrical bis-Schiff bases of 1, 6-hexanediamine: Synthesis, characterization, and in silico prediction of potential inhibitor against coronavirus (SARS-CoV- 2). Applied Organometallic Chemistry. 2021;**35**(1):e6067. DOI: 10.1002/aoc.6067

[62] Uddin MN, Chowdhury DA, Rony MM. Complexes of Schiff bases derived from 2-hydroxyaldehyde and propane-1,2-diamine: Synthesis, characterization and antibacterial screening. American Journal of Applied Chemistry. 2014;**1**:12-18. ISSN: 2381-4535

[63] Chowdhury DA, Uddin MN, Chowdhury GK. Synthesis and characterization of some dioxomolybdenum(VI) complexes of Schiff bases with ONNO donor ligands. The Chittagong University Journal of Science. 2006;**30**(1):85-88. ISSN 1561-1167

[64] Uddin MN, Salam MA, Siddique MAB. Complexes of VO^{3+} with dibasic tetradentate Schiff bases and their microbial studies. American Journal of Chemistry and Applications. 2014;**1**(2):19-27. ISSN Online: 2381-4535

[65] Uddin MN, Rahman S. An *in-silico* evaluation of some Schiff bases for their potency against SARS-CoV-2 Main protease, PASS prediction and ADMET studies. The Journal of Engineering and Applied Science. 2023;**10**(1):29-48